Mission Mathematics

Grades K–6

Mary Ellen Hynes

Contributors

Nancy Belsky   Catherine Blair
Eva Farley   Donn Hicks
Molly Ketterer   Barbara R. Morgan
Andrea Prejean   Terry Thode

Project Director

Michael C. Hynes

PUBLISHED IN COOPERATION WITH
THE NATIONAL AERONAUTICS AND
SPACE ADMINISTRATION

NATIONAL COUNCIL OF TEACHERS OF MATHEMATICS

Copyright © 1997 by
THE NATIONAL COUNCIL OF TEACHERS OF MATHEMATICS, INC.
1906 Association Drive, Reston, VA 20191-1593
All rights reserved

Second printing 1998

Library of Congress Cataloging-in-Publication Data

Mission mathematics / [edited by] Mary Ellen Hynes ; contributors,
   Nancy Belsky ... [et al.].
      p.   cm.
   "Grades K–6."
   "Prepared in cooperation with the National Aeronautics and Space
Administration."
   Includes bibliographical references.
   ISBN 0-87353-434-4 (pbk.)
   1. Mathematics—Study and teaching (Elementary) 2. Astronautics—
Mathematics. I. Hynes, Mary Ellen. II Belsky, Nancy.
III. United States. National Aeronautics and Space Administration.
QA135.5.M543   1997
372.7—dc21   97-8167
                                                                                                  CIP

Cover photograph copyright 1996 by Andy Steere. The sky around the bright star Deneb, in the constellation Cygnus; 30-minute exposure

Text photographs, unless otherwise credited, are courtesy of NASA.

> The publications of the National Council of Teachers of Mathematics present a variety of viewpoints. The views expressed or implied in this publication, unless otherwise noted, should not be interpreted as official positions of the Council.

Printed in the United States of America

CONTENTS

Foreword iv

Preface v

Matrix of Activities and Mathematics Standards vii

Aeronautics 1

    An Introduction to Aeronautics Activities 3
    Covering an Airplane 5
    Long-Distance Airplanes 12
    Target Practice 18
    Location of Load 23
    Flight Direction 28
    Rescue-Mission Game 34
    The Air Show 41

Human Exploration and Development of Space 44

    An Introduction to Human Exploration and
        Development of Space (HEDS) Activities 46
    Fizzy-Tablet Rockets 47
    Scrumptious Veggie Shuttle 51
    Tiles and Tessellations 53
    Water, Water 57
    Living Areas 61
    Spheres in Space 66
    Destination: Space Station 70

Space Science 72

    An Introduction to Space Science Activities 73
    Probing the Planets 74
    How Much Does the Milky Weigh? 79
    Gravity and Weight 82
    Mission to Mars 85
    Journey to Jupiter 89

Mission to Planet Earth 94

    An Introduction to Mission to Planet Earth Activities 96
    Protractor Rocket Launches 97
    Solar Observations Over Time 101
    Collecting the Rays 106
    Does the Sun Heat Fairly? 110
    Weather Watchers 113

Appendix A: NASA Resources for Educators 117

Appendix B: Charting the Planets 119

Bibliography 120

FOREWORD

Mission Mathematics: Linking Aerospace and the NCTM Standards is a collaborative project of the National Aeronautics and Space Administration (NASA) and the National Council of Teachers of Mathematics (NCTM). The vision of Frank Owens, director of educational programs for NASA, and James Gates, executive director of NCTM at the time of the conception of this project, inspired this unprecedented effort to link the science of aeronautics to the standards NCTM has developed for all aspects of mathematics education. The guidance of the project was entrusted to Pamela Mountjoy (NASA), Cynthia Rosso (NCTM), and Marilyn Hala (NCTM). Their support and expertise have facilitated the completion of the project, and their contributions have added immensely to the quality of the product.

The products of this project were developed using a writing team and a leader for that team. The editor for the grades K–6 book, Mary Ellen Hynes, deserves a lion's share of the credit for the quality of this book. Her efforts in undertaking the research for scientific accuracy, writing the activities, incorporating feedback from pilot-test results, compiling the activities, and editing the final manuscript have given the structure to this book. Selecting activities from the many exciting lessons submitted by the writing team was not an easy task. The writing team for this book consisted of Mary Ellen Hynes (editor), Nancy Belsky, Cathy Blair, Eva Farley, Donn Hicks, Molly Ketterer, Barbara Morgan, Andrea Prejean, and Terry Thode. The dedication that these educators have shown in the development of this book is phenomenal. As you use the book, I am sure that you will sense the enthusiasm each writing-team member had for using aerospace activities in the classroom to motivate students both to learn mathematics and to develop mathematical ideas.

The development of the activities for this project included pilot tests in classrooms throughout the nation. NCTM published a call in the *NCTM News Bulletin* for teacher-participants in this phase of the project. We could never have anticipated the response from classroom teachers. Our e-mail was overloaded, and the regular mail always seemed to have another offer to volunteer time for the project. Many teachers phoned us to follow up on their earlier correspondence. Clearly, our teachers are interested in using the context of aerospace to improve learning in their classrooms. Unfortunately, because of budget constraints in the project, we could select only a few teacher-participants from the many volunteers. The cost of duplicating the pilot materials and the postage to send classroom sets back and forth between teachers and the writing team prohibited us from using everyone who volunteered. For the pilot, classrooms were selected to give us a broad geographical, ethnic, and socioeconomic view of the effectiveness of the lessons. The large number of volunteers gave us the opportunity to reach a wide range of students. For those teachers who participated by sending feedback from their classroom experience with draft activities, thank you. Your feedback allowed the writing teams to make the activities more effective for children. Also, we thank all of you who volunteered to assist in the project. This type of spirit makes the teaching profession very rewarding.

In addition, we would like to acknowledge the NASA staff who provided us with background information and materials.

As can be seen, this book was the product of many hands and minds. This collaborative effort has been the engine that has made this important book a reality. Thank you all.

Michael C. Hynes, Project Director

PREFACE

The writing team created this book of activities with the following goals:

- To devise mathematical problems and tasks that focus on the NCTM curriculum and evaluation standards in the context of aerospace activities

- To engage students actively in NCTM's four process standards: problem solving, mathematical reasoning, communicating mathematics, and making mathematical connections (a) among topics in mathematics, (b) to other disciplines, and (c) to real life

- To translate the work of engineers and scientists at NASA into language and experiences appropriate for young learners

- To provide teachers with mathematics activities that can complement many of the available NASA resources for students and educators

The activities are grouped according to the missions that NASA has identified as its focus: Aeronautics, Human Exploration and Development of Space, Space Science, and Mission to Planet Earth. This organization can help you coordinate these mathematics activities with the many free materials available at the NASA Educators Resource Centers.

This book was developed using NCTM's mathematics standards for grades K–4. When you compare the K–4 mathematics standards with the 5–8 standards, you see a natural transition from one set to the other. Because the organization of many elementary schools includes K–5 or K–6 grades, fifth and sixth grades are included in the span of grades listed at the beginning of each Mission Mathematics activity.

The blue portion of the K–6 continuum identifies the range of the grade levels at which the students are most likely to experience success. Because it is a range, you may want to make appropriate adaptations of the activities for students at both ends of the range. Suggestions for modifying activities for younger or older students, cross references to related Mission Mathematics activities, and suggestions for further explorations are included.

You will frequently find the heading "Class Conversation," followed by a set of suggested questions. For the teacher, these questions motivate students, encourage students to reason and communicate, and offer a means to assess students' readiness or progress. For the students, a class conversation helps them relate previous experiences to the activity, focus on the mathematical task or problem, identify and explore patterns and relationships, apply a problem-solving strategy, and discover connections to other mathematical concepts or other cross-disciplinary links.

Many of the activities can be used as miniunits or investigations. Several tasks and problems are found within an activity. Each teacher must decide what is an appropriate amount of time for his or her class to work on these tasks and problems. Some activities can be used over several weeks.

To emphasize the anticipated role of young learners as they become engaged in these problem-solving activities, the word *students*, rather than the word *children*, is used throughout the book.

When initial phone calls were made to the NCTM members selected to be a part of the grades K–6 writing team of the Mission Mathematics project, no one hesitated—all were delighted to accept the invitation, the responsibilities, and the challenges. Their enthusiasm for, and contributions to, the project have continued to grow from that day. A true community of learners has emerged as they shared their ideas willingly and stimulated and supported one another.

The writing team members gave many, many hours to this project. In the process, they have formed new professional friendships and learned new ways of contributing to their own professional settings. The editor is grateful for their hard work, creative ideas, enthusiasm, and support throughout the project.

A special note of thanks is given to Cynthia Rosso, Marilyn Hala, and the production and editorial staff of NCTM; to Pamela Mountjoy of NASA; and to Michael Hynes, the director of the Mission Mathematics project. Each provided the support, guidance, and expertise that have made the project enjoyable from day one.

I hope that each of you will catch the aerospace fever and find several activities that you will enjoy sharing with your students.

TABLE OF NCTM MATHEMATICS STANDARDS AND AEROSPACE ACTIVITIES

NCTM Mathematics Standards

1. Estimation
2. Number Sense and Numeration
3. Whole Number Operations
4. Geometry and Spatial Sense
5. Measurement
6. Statistics and Probability
7. Fractions and Decimals
8. Patterns and Relationships

Activity	1	2	3	4	5	6	7	8
Aeronautics								
Covering an Airplane		x	x	x	x	x	x	x
Long-Distance Airplanes		x			x	x		x
Target Practice		x	x	x	x	x	x	x
Location of Load		x	x		x	x	x	x
Flight Direction		x			x	x	x	x
Rescue-Mission Game		x		x		x		x
The Air Show		x			x	x		x
Human Exploration and Development of Space								
Fizzy-Tablet Rockets		x			x	x	x	x
Scrumptious Veggie Shuttle		x			x			x
Tiles and Tessellations				x				x
Water, Water	x	x	x	x	x	x		x
Living Areas	x				x	x		x
Spheres in Space	x			x	x			x
Destination: Space Station		x		x				x
Space Science								
Probing the Planets	x	x			x	x	x	x
How Much Does the Milky Weigh?	x	x			x			x
Gravity and Weight	x	x			x	x	x	x
Mission to Mars						x		x
Journey to Jupiter						x		x
Mission to Planet Earth								
Protractor Rocket Launches	x	x			x	x		x
Solar Observations over Time				x	x	x		x
Collecting the Rays	x	x		x	x	x		x
Does the Sun Heat Fairly?		x			x	x		x
Weather Watchers		x			x	x		x

High-Speed Civil Transport

AERONAUTICS

When we watch newscasts of early attempts to fly, we get a sense of the adventure that early pioneers in flight experienced. Watching their funny-looking machines that had few successes and spectacular failures leads us to believe that manned flight was built haphazardly. Well, this belief has some truth to it, but these pioneers did engage in deliberate investigations. From these early endeavors was born the science of flight called *aeronautics*. Over the years, this science has grown in importance for our country. Today, aeronautics is vital to our national security and economic well-being.

For more than seventy-five years, the National Aeronautics and Space Administration (NASA) and its predecessor, the National Advisory Committee for Aeronautics, have worked closely with industry to give the United States a preeminent position in aeronautics. The aeronautics industry involves more than 1 million high-quality jobs, over $40 billion in exports annually, and almost $30 billion in positive balance of trade. Aeronautics, like other industries, is being challenged by international competitors, however. We cannot rest on our laurels. The challenge to improve aeronautical research is constant.

One area in which research is ongoing is the development of new planes for our commercial fleet so we can maintain our competitive edge in the industry. For example, the emergence of the Pacific Rim countries as financial powers makes the development of a next-generation supersonic transport a necessity. It is anticipated that the marketplace for high-speed commercial travel could be worth as much as $200 million annually and produce 140 000 jobs.

NASA supports research that addresses concerns about technical and environmental challenges related to supersonic travel. One problem involves the development of technology to ensure that emissions from supersonic jets do not damage the ozone layer. Another potential environmental insult is the sonic boom associated with supersonic flight. NASA scientists are working on wing and fuselage designs that will reduce the boom-effect levels without causing a significant loss in performance of the airplane.

Advanced helicopter and tilt-rotor craft could greatly reduce the air traffic at overcrowded major airports by allowing commuters to fly in and out of highly convenient vertical-takeoff-and-landing airports located in city centers. NASA is studying ways to reduce the potential adverse effects of these new aircraft. It is studying ways to reduce takeoff and landing noise as well as to improve the noise factor in passenger cabins.

Safety in the air for all commercial flight is a focus of NASA research. NASA scientists have produced designs, materials, and practices that have solved some of the safety problems related to commercial flight. As early as 1946, research efforts to solve the problem of icing on airplanes began. Ongoing research focuses on the application of deicing techniques to new airplanes under new conditions. We have learned about sudden changes in wind speed and direction found in wind shear and in storms called microbursts. NASA is researching ways to detect conditions for these events and to give pilots warning so they can avoid entering a dangerous environment. Additionally, NASA supports research to improve the materials used to build airliners, improve systems involved in takeoffs and landings, and consider human factors that may affect flight performance and safety.

NASA is a primary support system for aeronautics. From basic discoveries in research laboratories to advances in the design and manufacture of products to a concern for aircraft safety, NASA is a leading player in this important industry.

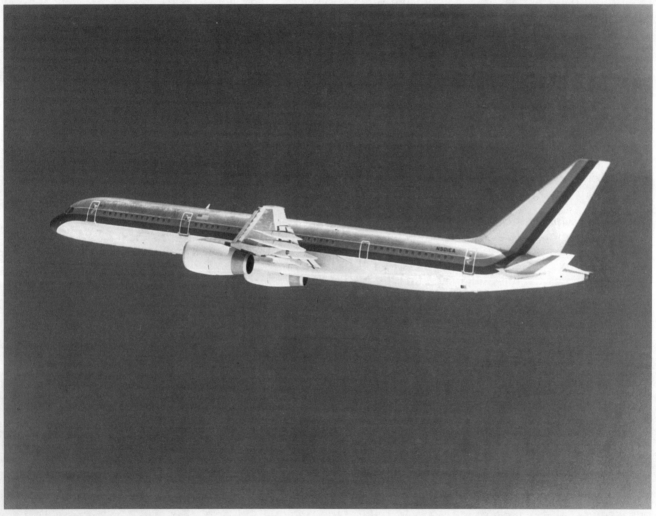

Boeing 757

AN INTRODUCTION TO AERONAUTICS ACTIVITIES

All activities in this section integrate the four NCTM process standards: problem solving, mathematical reasoning, mathematical communication, and mathematical connections. Each activity also incorporates several of the NCTM mathematics standards.

Many of the activities involve repeated test flights, which are necessary to give students adequate data to display in graphs and to analyze. Because the activities involve action with airplane test flights, participants should wear safety goggles.

The following brief outline describes the "Aeronautics" activities and the associated mathematics.

Activity Name	Mathematics	Grade Levels
Covering an Airplane	Area, picture graph, pictograph, bar graph	K–6
Long-Distance Airplanes	Measuring distance, range, median, mode, stem-and-leaf plot	2–6
Target Practice	Mean, line plot	2–6
Location of Load	Measuring distance, multiple-bar graph	3–6
Flight Direction	Fractions and decimals, circle graph	3–6
Rescue-Mission Game	Plotting points on a coordinate grid, probability	3–6
The Air Show	Measuring distance and time, glyphs	K–6

The activities in this section can help students learn about the aeronautics activities of NASA. A good visual image that conveys NASA's commitments to aeronautics and space exploration is a Space Shuttle riding piggyback on a modified Boeing 747 airplane.

The National Aeronautics and Space Administration is known worldwide for its exploration of space. However, as the name of the agency clearly states, aeronautics has always been a major component of NASA's missions. Before NASA, came NACA. In 1958, the National Advisory Committee for Aeronautics (NACA) became NASA.

NASA's aeronautical knowledge and products are used not only in experimental research aircraft but also in commercial and general-aviation aircraft. NASA continues aeronautical research in such programs as human interaction with automation and aerospace hardware; wind-tunnel research; the development of thermodynamics, materials, and advanced engines; and nondestructive evaluation and inspection methods.

NASA's newer generations of research aircraft provide data to prepare for a High-Speed Civil Transport (HSCT).

NASA Aeronautics Centers are located at NASA Ames Research Center, NASA Lewis Research Center, NASA Langley Research Center, and NASA Dryden Flight Research Center. The NASA Educators Resource Centers at these locations have information on aeronautics for teachers and students. Other sources of aeronautics information are available through NASA's home page on the Internet. For more information about NASA's materials for teachers, see appendix A.

X-29

COVERING AN AIRPLANE   K 1 2 3 4 5 6

In this problem-solving lesson, students cover the surface area of an airplane using small squares (1 in.²), large squares (4 in.²), rectangles (2 in.²), and right triangles (1/2 in.²). They use tally marks to record the number of each shape used and create a picture graph, a pictograph, or a bar graph to represent their data. Finally, they find the measure of the area in square inches using counting, addition, and multiplication.

PURPOSES

- To cover and measure an area using a set of given shapes
- To collect, organize, and interpret data
- To construct a picture graph, a pictograph, and a bar graph
- To use a standard unit of measure for area, 1 square inch

INTRODUCTION

To help students understand NASA's role in the aeronautics industry, discuss some of the important design and construction features of airplanes. All parts must be carefully measured so that they fit together exactly. The surface area of the frame must be carefully measured to ensure complete coverage with materials that can withstand extreme changes of temperature; severe weather conditions, such as rain, hail, and snow; and changes of atmospheric pressure. NASA engineers test new designs and new materials for covering aircraft.

GETTING STARTED

Invite your students to pretend that they are NASA engineers who must carefully measure their airplane's surface using a set of special tools—squares, rectangles, and right triangles.

Ask students to identify and describe the outline of an airplane as you distribute resource page 9.

Class Conversation

- Why do most airplanes have this shape?
- What other smaller shapes do you see within the airplane? Where?
- What shapes can we use to cover the airplane completely?

DEVELOPING THE ACTIVITY

After distributing resource page 10, ask students to identify the different shapes they see. Review the characteristics of a square, a right triangle, and a rectangle. Have students cut out all the shapes. Ask how the shapes are related to one another. Help students see equivalence relations among the areas of these shapes—two small squares cover a rectangle, four small squares cover a large square, two triangles cover a small square, and so on. Students should work with a partner to solve the following problem.

> Pretend that you are NASA engineers. You need to cover this airplane with a new material. But first, you must measure the airplane using a special set of tools to see how much material you need.
>
> Use your shapes to cover as much of the airplane as you can. You can turn your shapes around or flip them over to arrange them in any way you want, but the shapes must stay inside the outline of the airplane.

NCTM Mathematics Standards
- **Geometry and Spatial Sense**
- **Statistics**
- **Measurement**
- **Fractions and Decimals**
- **Patterns and Relationships**
- **Number Sense and Numeration**
- **Whole Number Operations**

Materials: *Resource pages 9, 10, and 11, transparencies of resource pages 9, 10, and 11, glue, scissors, crayons, pencils*

Note for Primary-Grades Teachers: *This activity can be adapted for students who are not ready for experiences with standard units of measure. Limit the activity to covering an area with the different geometric shapes and then counting the shapes. Simply rephrase the problem so students find the number of each type of shape used to cover the airplane. Present graphing tasks and methods appropriate for your students' experience levels.*

When students have completely covered their airplanes, ask them to glue or tape the pieces inside the outline of the airplane. Allow enough time for students to share their airplanes and their different strategies for covering their planes. Then use the following questions to help students discuss their results.

Class Conversation

- Which shapes did you use to cover the airplane?
- Did you use more squares or rectangles?
- Did you use more triangles or squares?

Ask students to think about how they answered these questions. Most will respond that they looked at their planes and counted the pieces. Ask if they can think of another way to answer these questions without having to count each time. Since different ways to cover the airplane are possible, talk about the need to organize and display their information to help them solve the problem.

As appropriate for your students, review or introduce the procedure for recording tally marks in a table to show the number of small and large squares, rectangles, and triangles used to cover their airplane. After they have completed a tally table, ask if they would like to display their data in graphs. Use this opportunity to review or introduce any or all of the following types of graphs.

A *picture graph* uses a one-to-one correspondence between the shapes covering the airplane and the shapes shown in the graph. For example, if the student used eight triangles when covering the airplane, eight triangles would be shown in the graph. Students count by ones.

A *pictograph* uses a many-to-one correspondence between the shapes covering the airplane and the shapes shown in the graph. For example, if the student used a two-to-one correspondence and twelve small squares to cover the airplane, six small squares would be shown in the graph. Students count by twos and may discover the difference in trying to graph even and odd numbers. Pictographs include a key to indicate the value of each figure.

A *bar graph* uses a scale to show the frequency for each of the given figures. The scale is placed along the vertical axis for a vertical bar graph, as on resource page 11, or along the horizontal axis for a horizontal bar graph. Help students select a scale appropriate for the range of the data.

Ask a series of questions to help students see how useful their tables and graphs are for solving problems. Emphasize that their answers are not necessarily the same, since they used different ways of covering their airplanes.

Class Conversation

- Which shape was used most to cover your airplane?
- Did you use more than five right triangles?
- Did you use fewer than ten small squares?
- Which shapes did you use an even number of times? An odd number of times?
- What is the total number of shapes you used?
- How does your graph make it easy to answer questions like these?

Students should share their graphs with the class and summarize what they notice about their data while using their graphs. Invite students to compare their graph with the graphs of their classmates. Ask questions that will help them see that their graphs are not necessarily the same.

NCTM Teaching Standards: *As students share with their classmates, reinforce the idea that it is good to see so many different strategies and results and that there are many correct ways to solve this problem. Respecting and valuing students' ideas and ways of thinking are important to a good learning environment.*

Technology Connection: *If graphing software is available, have students use it to graph this set of data.*

Standard Units of Measure

Remind students that they still need to answer the question, How much material do you need to cover your airplane? Ask how they can use the information in their graphs or table to measure the amount of material needed to cover the airplane. Introduce or review the need for standard units of measure. Talk about a unit that is used to measure area—the square inch.

If appropriate for your students, use a ruler to show that each side of the small square measures 1 inch. Tell them that this area is named *1 square inch*. Then ask students to compare each of the other shapes with the small square. Help them see that the area of the rectangle is equivalent to two small squares and will cover an area of 2 square inches. Likewise, the area of the large square covers an area of 4 square inches, and the area of the right triangle covers an area of 1/2 square inch.

After students have become familiar with the units of measure associated with the different shapes, pose the final problem for this activity.

> Use your table or graph to find how many square inches of material are needed to cover your airplane.

Problem Solving: *Students can use both the make-a-table strategy and the make-a-graph strategy in this investigation.*

Students will be able to solve the problem using a variety of strategies, such as counting and adding the measures. To find the area covered by the right triangles, many students will group two triangles together to form 1-inch squares, which can be a beginning for a discussion related to the addition of fractional numbers.

Some students may be ready to use multiplication to find the areas covered by each of the given shapes. For example, if students use eight small squares, five rectangles, six triangles, and three large squares, they could use the following number sentences to find the area of the airplane in square inches:

$$\left. \begin{array}{l} 8 \times 1 = 8 \\ 5 \times 2 = 10 \\ \frac{1}{2}+\frac{1}{2}+\frac{1}{2}+\frac{1}{2}+\frac{1}{2}+\frac{1}{2} = 3 \\ 3 \times 4 = 12 \end{array} \right\} \quad 8 + 10 + 3 + 12 = 33$$

CLOSING THE ACTIVITY

As they share their solutions with their classmates, help students record in a line plot the number of square inches they used to cover the airplane. A sample line plot is shown on the next page. Students should see that even though the airplanes were covered with different sets of shapes and that all the graphs were not the same, everyone reached approximately the same measure in square inches.

As appropriate in your class, have students summarize their activities and their findings.

NCTM Assessment Standards: *To make assessment a coherent process, teachers should use assessment activities appropriate to the investigation. In this set of activities, there are several opportunities for students to develop products (graphs, generalizations, computations, pictures). A compilation of these can become a miniportfolio.*

FURTHER EXPLORATIONS

Give students a piece of 1-inch graph paper and have them draw a horizontal axis and a vertical axis intersecting near the lower-left corner. Ask students to number the horizontal axis from 1 through 8 and the vertical axis from 1 through 10. Ask students to trace their airplane on this sheet and then give the ordered pairs for the endpoints of the line segments.

Covering an Airplane

Sample Line Plot for Number of Square Inches to Cover the Airplane

	31	32	33	34	35	36	37	38
9					x			
8					x			
7					x	x		
6				x	x	x		
5				x	x	x		
4				x	x	x		
3			x	x	x	x	x	
2		x	x	x	x	x	x	
1	x	x	x	x	x	x	x	x
Square Inches	31	32	33	34	35	36	37	38

RELATED ACTIVITIES IN MISSION MATHEMATICS

Another activity related to covering an area is "Tiles and Tessellations," which focuses on motion geometry, slides, flips, and turns rather than on standard units of measure for area.

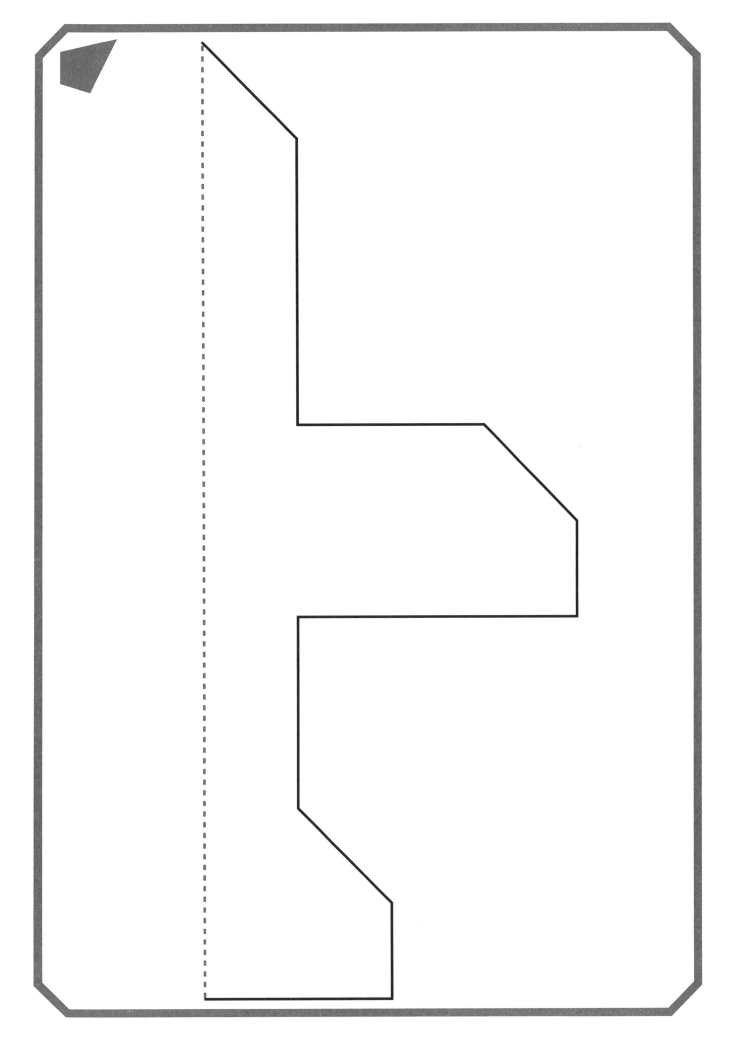

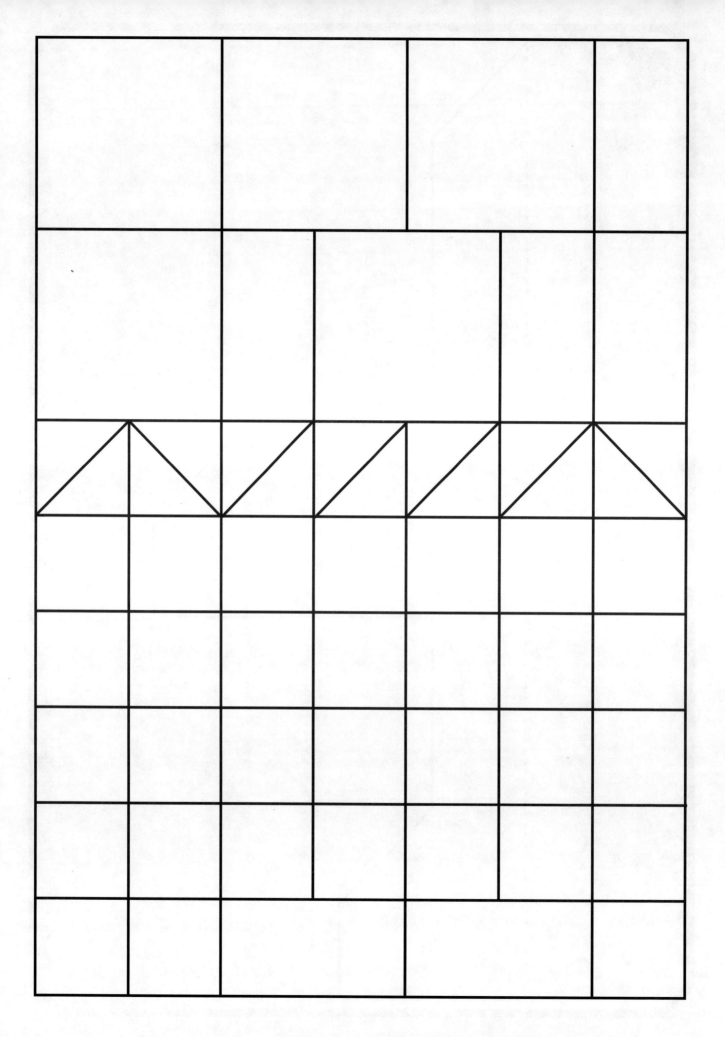

BAR GRAPH

_____
(Title)

16
15
14
13
12
11
10
9
8
7
6
5
4
3
2
1
0

What I Discovered
_____
_____
_____
_____

Covering an Airplane

LONG-DISTANCE AIRPLANES   K 1 2 3 4 5 6

NCTM Mathematics Standards
- *Statistics and Probability*
- *Number Sense*
- *Patterns and Relationships*
- *Measurement*

Materials for each student: *Resource pages 16 and 17, 1 sheet of paper*

Materials for each pair or group: *rulers, yardsticks, or measuring tape*

Management Tip: *When planning this activity, consider (1) the availability of a large space with no wind, such as the cafeteria or a long hallway, and (2) the number of sessions needed to conduct test flights and collect data.*

Students make paper airplanes and explore attributes related to increasing flight distances. Each student collects data from three flights of the airplane and finds the median distance. Students then collect, organize, display, and interpret the median distances for the class in a stem-and-leaf plot.

PURPOSES
- To design and construct an airplane to fly long distances
- To collect, organize, display, and interpret data
- To determine median distances
- To construct a stem-and-leaf plot of class data
- To determine the range and mode of class data

INTRODUCTION
To introduce this lesson and help students understand some of NASA's roles in the aeronautics industry, talk about NASA engineers and their work. One challenge for aeronautical engineering is to design planes that can fly long distances while carrying large numbers of people or heavy loads of cargo. Australia and China are two examples of distant destinations where both people and cargo need to go. If the United States can build the best planes to fly to such far-away destinations, we will benefit from increased trade with these countries. Designing airplanes to fly long distances is serious business.

GETTING STARTED
Review the forces that affect flight: lift, gravity, thrust, and drag (air resistance). For information about these four forces, see the activity "Rescue-Mission Game."

Ask students to pretend that they are aeronautical engineers for NASA. Their mission is to design and construct a paper airplane that will travel the greatest distance.

DEVELOPING THE ACTIVITY
As students make different airplanes to explore the attributes that may affect the distances their airplanes fly, introduce or review the concept of symmetry. Encourage students to observe and analyze their classmates' airplanes.

Class Conversation
- Which airplanes are similar to yours?
- What attributes do they have in common?
- Which airplanes do you think will travel the same distances as your airplane? Why?

After students record their predictions on the data sheet, they should fly their airplanes three times, using the same amount of force each time.

Students should work with a partner or in a small group to measure their flight distances in inches and record all three distances on a data-collection sheet. To find their median, or middle, distance, students should follow the directions on the flight data sheet, resource page 16.

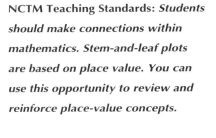

MAKING A STEM-AND-LEAF PLOT

Have students write their median flight distances on the chalkboard in an orderly, but random, way. The following is a sample set of data with the corresponding stem-and-leaf plot.

Sample Data: Median Flight Distances (in inches)

72 74 49 73 92 75 73 82 114 52 82 123 81
67 101 70 76 73 96 73 43 76   75 108 74   64
48 81 120 80 57 68 105 76 73   41 112 77   69

NCTM Teaching Standards: *Students should make connections within mathematics. Stem-and-leaf plots are based on place value. You can use this opportunity to review and reinforce place-value concepts.*

Stem-and-Leaf Plot for Median Flight Distances (in inches)

Stems	Leaves
4	1 3 8 9
5	2 7
6	4 7 8 9
7	0 2 3 3 3 3 3 4 4 5 5 6 6 6 7
8	0 1 1 2 2
9	2 6
10	1 5 8
11	2 4
12	0 3

6|4 represents 64

As you record each leaf value in a row of data, cross off the corresponding data point from the original set of data. For example, in the first row of this stem-and-leaf plot, after recording the 1, cross off the 41 in the data set. After recording the 3 in the first row, cross off the 43, and so on.

Class Conversation

- What number should I cross off after I record the 2 in the second row of our stem-and-leaf plot? After I record the 7?
- In the third row, what does each number represent?
- In what row is the distance of 105 inches recorded?

Continue with this type of question until students recognize and describe the pattern for the organization.

- How are the data grouped or organized?
- What does each row of the graph represent?
- Why do you think that this graph is called a stem-and-leaf plot? What numbers are the stems? The leaves?
- How does this graph help us "see" the data better?
- Do you think that this method is the best way to organize the data? Why or why not?

After students complete the class stem-and-leaf plot, ask them to line up the airplanes in order from the one that had the least median flight distance to the one with the greatest median distance. Use this display to discuss the *range* of flight measurements.

If this type of graph is not new to your students, invite them to participate in setting up the stem-and-leaf plot. Ask students what values would be best for the stems. Different responses should be anticipated, depending on their flight results.

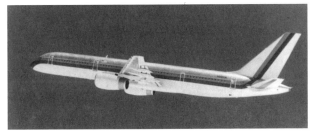

Boeing 757

To reduce drag, NASA, the Air Force, and Boeing are researching a suction system on wings to siphon off turbulent air through millions of tiny laser-drilled holes. It is called laminar flow control and could result in major fuel savings in commercial aircraft. Additional NASA connections and background information are in the introduction for the "Aeronautics" section.

Instructional Note: *Help students see that the tens place of their measurements is represented by the stem in the appropriate row and is recorded to the left of the vertical bar. The leaves represent the ones place and are recorded to the right. All entries in any row represent data points in the same decade. For example, any entries in the row for the stem 18 would be between 180 and 189 inclusive.*

Encourage students to use the range of their data to determine what values are needed for the stems. For example, if all distances are 86 inches or more, but less than 240 inches, they may suggest a graph with stems like the sample shown below.

Stem-and-Leaf Plot for Median Distances (in inches)

Stems	Leaves
8	
9	
10	
11	
12	
13	
14	
15	
16	
17	
18	
19	
20	
21	
22	
23	

CLOSING THE ACTIVITY

This activity can reinforce important place-value concepts. Many students have renamed 21 dimes as 210 cents or $2.10 but have not had opportunities to see that this same type of renaming of numbers can be done in other applications, such as with units of linear measure. This activity can contribute to students' understandings of patterns and relations among numbers in our decimal number system.

Class Conversation

- What patterns or clusters do you see in our data?
- Did other airplanes have about the same median distance as yours?
- What similarities do other airplanes have with yours?
- Did your predictions match your results?
- How can we find the *median* distance for our whole class?
- What is the *mode,* or most frequently occurring, distance in our graph?
- Look at the airplanes that traveled the greatest and least distances. How are they alike? Different?

ASSESSING THE ACTIVITY

NCTM Assessment Standards: *Teachers should use a variety of assessment techniques to collect data that can inform teachers as they make instructional decisions. Remind students that researchers for NASA always write a report with their conclusions from their experiments. If appropriate for your students, ask them to write their conclusions about the activity.*

This activity provides a variety of assessment opportunities. In their discussions, encourage students to—

- share their individual conclusions with the class;
- summarize the class activities: they made predictions, flew airplanes three times, determined median distances, displayed data in a stem-and-leaf plot, used the graph to draw conclusions about their experiment and data, and recorded their conclusions;
- identify the median, mode, and range of their class data set;
- explain what each row of numbers in the class stem-and-leaf plot represents.

As students use the stem-and-leaf plots, check that students' plots contain all data points in the correct rows and have an appropriate title. As the activity ends, check that students make and record appropriate concluding remarks about the class activity and data.

FURTHER EXPLORATIONS

- Students can decorate and identify their airplanes with their initials and flight distances. Share the following aviation information with them. All airplanes have a letter and number combination painted on their surfaces for identification purposes. Each set of letters and numbers is registered and belongs to that airplane only. The letters and numbers are called in by the pilot to identify the airplane to airport control towers and base operators when requesting airport information for landing. Other aircraft in the vicinity are also able to hear the call and be aware of the airplane's presence. The identification letters are called in by using the phonetic alphabet. Give students a copy of this alphabet. The numbers are called in one at a time, with only one change. "Nine" is pronounced "niner."

- Students can turn their papers 90 degrees counterclockwise so their stem-and-leaf plots are facing sideways. Ask these questions:

    — What do you notice about our graph when we turn our display sideways? Does it remind you of any other type of graph with which you have worked before?

    — What similarities do you notice between the stem-and-leaf plot and the bar graph? What differences?

    — Does holding your stem-and-leaf plot sideways help you see the results more easily? Explain.

While students have their stem-and-leaf plots turned 90 degrees, have them place a piece of white paper over them and trace the shape. They can use the traced shape to make a bar graph. If they draw the bars adjacent to one another with no space in between, they have drawn a *histogram*. Histograms show data grouped in equal intervals. Remind students to give their graphs a title and to label the vertical and horizontal axes. Encourage a discussion of their bar graphs or histograms.

RELATED ACTIVITIES IN MISSION MATHEMATICS

"Rescue-Mission Game" provides background information about the forces of flight. "Flight Direction," "Target Practice," and "Location of Load" are all related to experimenting with airplanes.

A: Alpha	O: Oscar
B: Bravo	P: Papa
C: Charlie	Q: Quebec
D: Delta	(kaybec)
E: Echo	R: Romeo
F: Foxtrot	S: Sierra
G: Golf	T: Tango
H: Hotel	U: Uniform
I: India	V: Victor
J: Juliet	W: Whiskey
K: Kilo	X: X-ray
L: Lima (leema)	Y: Yankee
M: Mike	Z: Zulu
N: November	

Instructional Note: *If your students have had several experiences with stem-and-leaf plots, they may be ready to learn when it is best to use each type of graph. The stem-and-leaf plot shows the total amount in each category and shows each individual score, whereas the bar graph shows totals only.*

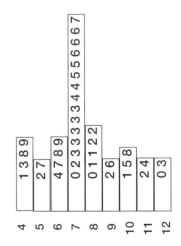

Sample histogram

Long-Distance Airplanes

FLIGHT DATA SHEET

Prediction: I think my airplane will fly about as far as

_____'s airplane because _____

_____

_____

_____

My Flight Data

Flight 1	Flight 2	Flight 3
_____	_____	_____
(distance in inches)	(distance in inches)	(distance in inches)

Steps for Finding My Median Flight Distance

1. Put an X through the greatest distance.

2. Put an X through the least distance.

3. Circle the remaining distance.

4. Write it here. _____ This distance is named the *median* flight distance.

Stem-and-Leaf Plot
Median Distances (in Inches) for Our Class

Stems	Leaves
0	
1	
2	
3	
4	
5	
6	
7	
8	
9	
10	
11	
12	
13	
14	
15	
16	
17	
18	
19	
20	
21	
22	
23	
24	

Conclusions _____

_____

_____

_____

_____

Long-Distance Airplanes

TARGET PRACTICE

K 1 2 3 4 5 6

Students make paper airplanes that fly accurately and land in a designated target area. They conduct ten test flights and record the number of airplanes that successfully land in the target. Students model and graph their class data and find the mean value of the data.

NCTM Mathematics Standards
- *Statistics*
- *Patterns and Relationships*
- *Geometry and Spatial Sense*
- *Measurement*
- *Number Sense and Numeration*
- *Fractions and Decimals*
- *Whole Number Operations*

Materials: *Paper, clothesline or jump rope long enough to form a target circle with a diameter of approximately 6 feet, yardsticks, string, resource page 22*

PURPOSES

- To make a paper airplane that can fly into a target area
- To collect, organize, display, and interpret data
- To model the mean of a data set
- To display class data on a line plot

NASA tests experimental aircraft in the world's largest wind tunnels at Ames Research Center in California. Other NASA connections and background information related to the dynamics of flight and aeronautics can be found in the introduction to the "Aeronautics" section.

Note to Primary-Grades Teachers:
For younger students, you may want to work with small groups of about five students and limit data collection to five test flights.

Depending on the size of your students, you may want to adjust the size of the target area and the distance that students stand from the target.

GETTING STARTED

After sharing appropriate background information about NASA and aeronautics, invite students to pretend that they are NASA engineers and pilots working on a project.

Class Conversation

- Think about airplanes, helicopters, and Space Shuttles. Why is it important for a pilot to be in control of where he or she flies and lands the aircraft?
- How do you think that NASA engineers who design aircraft do experiments to make aircraft land accurately?
- What things should we think about to make our paper airplanes fly accurately?

Challenge your students to experiment with and make paper airplanes that will fly accurately and land in a designated target area. Then present the following task:

You are to design and make paper airplanes that can land accurately inside a target circle. We will do test flights and record how many of our airplanes are successful in landing in our target area.

Before students make their airplanes, help them visualize the size of a circular target area with a 6-foot diameter and a distance of 15 feet.

To help students become familiar with these distances and make comparisons between the two measures, ask them to use a yardstick to measure and mark multiples of a yard on a long piece of string. Help your students see that the 6-foot distance is two lengths of the yardstick and the 15-foot distance is five lengths of the yardstick.

Prerequisite Measurement and Geometry Concepts: *Students should be able to—*

- *use the standard unit of measure, the foot;*
- *use a yardstick or tape measure;*
- *understand the words* circle *and* diameter.

DEVELOPING THE ACTIVITY
Making the Airplanes

Students use a sheet of paper to make an airplane that they think will be able to land in the target area when flown from the designated distance.

Preparing the Target Area

Once the airplanes are made, take the class to a place with minimal wind interference and enough space to make the target circle and fly the airplanes. Students should bring measuring tools, their airplanes, data-collection sheets, and pencils.

Ask the class to stand in a large circle. Inside the circle, a small group of students can use a yardstick or their string to measure the 6-foot diameter and the jump rope or clothesline to mark the circumference of the target circle. Then students should measure a distance of 15 feet outside the target circle. Students should stand at this distance to conduct the test flights.

Management Tips: *Organize small groups of students into ten data teams. Each team collects and models the data for one test flight.*

If activity time or attention spans are limited, conduct test flights over several days in a series of shorter sessions. If multiple days are used, fly airplanes for any absentees to keep the total number of airplanes flown the same for all test flights.

Collecting the Data

Before the first flight, have students write on their data-collection sheets their predictions of how many airplanes will land inside the target circle.

After the class conducts the first test flight, the first data team observes and reports the results for the whole class. As one student counts the total number of airplanes in the target circle, the class members can record tally marks on their data sheets. Another student can verify the count; if necessary, a third student can confirm the count. All students in the class should record the same number of successful landings on their data sheets in a row labeled Flight 1.

Then the same data team counts and verifies the number of airplanes that land outside the target area. All students should record the same number of tallies for the number of unsuccessful landings. The last number recorded in the row is the sum of the first two numbers, the total number of airplanes flown in flight 1. You may want students to wait to find this sum until after they return to the classroom.

Class Conversation

- Were your predictions close to our results? Too low? Too high?
- Would you like to make any small adjustments to your airplanes?

Students can quickly make adjustments to their airplanes and then begin the next test flight. Repeat this procedure for a total of ten test flights. Following the collection of data from the last flight, students return to the classroom to model, analyze, display, and interpret the data.

Target Practice

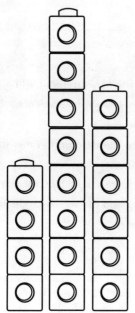

Sample display of data with connecting cubes

Manipulative Tips: *If connecting cubes are not available, use any type of uniformly sized object. Counters can be used, but squares, rectangles, or cubes work best because they fit together with no gaps. Index cards, color tiles, or other flat, uniformly sized pieces can be substituted.*

To emphasize groupings of 10, you may want to distribute connecting cubes or color tiles so there are 10 of each color. For example, if each team needs 30 cubes, give them 10 cubes in each of 3 colors.

Displaying the Data

Give each data team enough manipulatives, such as connecting cubes, to model the number of successful airplanes in their test flight. When each team has counted and connected the proper number of cubes for its model, display the models on the floor or a large table.

Ask students to use the models and their data-collection sheets to compare the results of the different test flights. In a class discussion, ask students to answer the following questions and tell how they found the answer. Did they use the model or the data-collection sheet?

Class Conversation

- Which test flight had the most successes? Which had the fewest? How do you know?
- What do you notice about the total number of airplanes flown on each test flight?
- In test flight 1, were more than half the airplanes successful? How do you know? What about the other test flights?
- What fraction can you write to show the number of successful airplanes in flight 1? In each of the other test flights?
- What do you notice about the denominators of these fractions? What do you notice about the numerators? Explain your thinking about your observations.
- Are the results for the test flights about the same or are they very different? How do our models help us see that result?
- Do you see any patterns or clusters in our test-flight data? Did you become more successful pilots as you had more practice and experience? How do you know?

Show students how to record the models of their test-flight data using a line plot such as this sample, then help them make a class line plot with the data they have collected.

Number of Successful Airplanes in Each Test Flight*

	1	2	3	4	5	6	7	8	9	10
16									x	
15									x	
14								x	x	
13								x	x	x
12						x		x	x	x
11					x	x	x	x	x	x
10				x	x	x	x	x	x	x
9	x			x	x	x	x	x	x	x
8	x	x	x	x	x	x	x	x	x	x
7	x	x	x	x	x	x	x	x	x	x
6	x	x	x	x	x	x	x	x	x	x
5	x	x	x	x	x	x	x	x	x	x
4	x	x	x	x	x	x	x	x	x	x
3	x	x	x	x	x	x	x	x	x	x
2	x	x	x	x	x	x	x	x	x	x
1	x	x	x	x	x	x	x	x	x	x
Test-Flight Number	1	2	3	4	5	6	7	8	9	10
	$\frac{9}{21}$	$\frac{8}{21}$	$\frac{9}{21}$	$\frac{10}{21}$	$\frac{11}{21}$	$\frac{12}{21}$	$\frac{11}{21}$	$\frac{14}{21}$	$\frac{16}{21}$	$\frac{13}{21}$

* Twenty-one planes were flown on each test flight.

Under each test-flight number, ask students to record a fraction that shows the part of the flights that were successful. This set of fractions should not be simplified. It is an example in which simplifying the fractions does not help in comparing the numbers. Seeing the same denominator for all the fractions is useful, since the denominator tells the students the total number of airplanes flown in each flight.

Discuss how the models, the line plot, and the fractions are different ways of showing their test-flight results.

Using Manipulatives to Find the Mean

Introduce the concept of the mean of a data set by asking students to think of just one number that they can use to tell their families about all the data collected by the class for the ten test flights.

Ask one member of each data team to come to its previously constructed model of the number of successful flights. Explain that the goal is to make each model have about the same number of cubes. Start with the test-flight model with the greatest number of cubes. Ask that team to remove one cube and add it to a model with fewer cubes. For example, in the sample data, the data team for flight 9 could take one cube off its model and add it to the model for flight 2. Data team members take turns repeating this process until the models for all, or nearly all, test flights have the same number of cubes.

At the end of this process in the sample data, the models for nine of the test flights would have ten cubes and the model for one flight would have nine cubes. The mean number for the sample data set is about 10.

CLOSING THE ACTIVITY

As the process of finding the mean number of successful airplanes for all test flights concludes, remind students that they did not throw away any of the cubes from the models; they just spread them out more evenly. They moved cubes from the models with more cubes to the models with fewer cubes until all models had about the same number of cubes.

FURTHER EXPLORATIONS

- Introduce the algorithm for computing the mean of a data set.
- As an alternative to using tally marks, students can use the following system. When they are finished, they just add a key and have a pictograph.

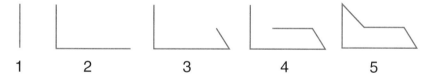

School-Home Connection

Students can collect new data for analysis with their families at a family math night. In presenting their class findings and conclusions, they learn how to prepare an oral presentation. In the process of rehearsing, they revisit the mathematics several times.

RELATED ACTIVITY IN MISSION MATHEMATICS

In a related activity, "Flight Direction," students make a circle graph.

Algebraic Thinking: *Experiences with multiple representations of numbers and sets of data help build important foundations in algebraic reasoning.*

Language Connection: *Ask students if they have heard or used the word* average *and what it means to them. Discuss that adults sometimes use the word* average *when they are really talking about the* mean.

NCTM Assessment Standards: *Throughout this activity are many opportunities to integrate assessment. An important assessment component is to ask students to summarize such items as the sequence of events in their experiment, their findings, and their conclusions.*

DATA-COLLECTION SHEET

Test Flight	Number of Successful Airplanes	Number of Unsuccessful Airplanes	Total
Flight 1			
Flight 2			
Flight 3			
Flight 4			
Flight 5			
Flight 6			
Flight 7			
Flight 8			
Flight 9			
Flight 10			

LOCATION OF LOAD   K 1 2 **3 4 5 6**

Students make and test paper airplanes with paper clips located in different positions. They measure the distances in meters, record their data in decimal notation, and display their data in a bar graph.

PURPOSES

- To explore the effect of load location
- To measure distances in meters
- To record distances as decimal numbers
- To collect, organize, and interpret data
- To construct a single-bar graph and a triple-bar graph
- To compare two different graphs

INTRODUCTION

Many of the research efforts of NASA's Office of Aeronautics involve improving the efficiency of airplanes. In this activity, your students can continue to pretend to be NASA scientists and engineers investigating how to improve the distance an airplane flies. The variable for this experiment is related to the center of gravity, the location of the cargo or load.

GETTING STARTED

Begin the activity with these questions:

- How many of you have ever been on an airplane? Were you seated in the front, middle, or back of the plane?
- Do you think that the plane would fly as well if all the passengers and cargo were located either at the back of the plane or at the front of the plane? Why?
- Why do you think that the people who design planes for NASA think about and discuss weight distribution?
- How can scientists and engineers do tests to see how the location of weight affects how well the plane flies?

DEVELOPING THE ACTIVITY

As you give each student a sheet of paper for making an airplane, four paper clips, and resource page 26 for recording data, present the following problem.

> First, each of you will design and make an airplane that you think will travel far. Next, we will experiment to find whether the position of a load affects how far your airplane flies.

Take the class to an open area with no wind interference to fly their airplanes. After an initial test flight with no loads attached, students should predict how far their airplanes will fly when the paper clips are positioned all at the front, all at the middle, and finally, all at the back of the airplane. Have them record their predictions.

For the front-load test, students put all four paper clips at the front of their airplanes and fly the airplanes. They measure the distance of the flight with metersticks or tape measures. Each student records his or her distance using decimal notation. With the paper clips in the same position, students repeat this process twice.

NCTM Mathematics Standards
- *Statistics*
- *Number Sense*
- *Measurement*
- *Patterns and Relationships*
- *Fractions and Decimals*
- *Number Operations*

Materials: *Paper, metersticks, crayons in three colors, resource pages 26 and 27, four paper clips for each student, graph paper (optional)*

When NASA engineers planned how to use an airliner to transport Space Shuttles to different NASA centers, they had to decide how to balance the load. Other NASA connections and background information about aircraft are in the introduction for the "Aeronautics" section.

Management Tip: *You may want to have three data-collecting sessions for the three test flights.*

NCTM Teaching Standards: *The learning environment of a class can be effective if students work collaboratively to make sense of mathematics. You may want to organize your class into flight crews for the test-flight portion of this activity. They can assist one another in measuring distances and recording data properly.*

NCTM Teaching Standards: *It is important that teachers decide when to use mathematical notations. Some students may not be ready to use decimal notation when measuring in meters. To reinforce whole-number concepts and operations, these students can measure distances in centimeters. For example, 5.23 m would be read and recorded as 523 cm.*

Technology Connection: *If it is available, students can use graphing software.*

Assessment Note: *Students can use their data-collection sheets and graphs as ongoing records of their experiment.*

For the middle-load test, students move all four paper clips to the middle of the airplane and measure and record the distances of three flights.

For the back-load test, students repeat this process with the paper clips at the back of the airplane.

When the data have been collected for all nine flights and students are ready to return to the classroom, ask them to think about the different graphs they know how to make. Is there one graph they can use to display all their data? After returning to the classroom, distribute the triple-bar-graph sheet on resource page 27 and explain to students that they are going to learn how to display all nine flights on one graph, a *triple-bar graph*. The graph has three bars for each of the three flights. Each bar will be colored with one of three colors to represent one of the three different locations of the paper clips.

Show students the box labeled "key" on their triple-bar-graph sheet. In the key, have students color the left figure red to represent front-loaded flights, the middle figure blue to represent middle-loaded flights, and the right figure green to represent back-loaded flights.

Help students use their data to construct a triple-bar graph similar to the sample shown below. This type of graph gives a visual representation of each flight of the experiment and, at the same time, allows students to compare the effects of load location.

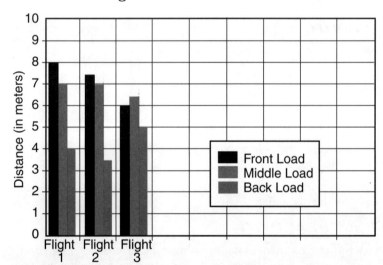

CLOSING THE ACTIVITY

Use the following questions to help students communicate their understandings of the experiment, the data collected, the graphical representation of their results, and their conclusions.

Class Conversation

- Does anyone have a graph on which the tallest bar is the same color in all three flights? What does this mean?

- Does anyone have a graph that does not show the same winner in all three flights?

- What color is the tallest bar on your graph in flight 1? In flight 2? In flight 3? What does this mean?

- What conclusions can we draw from our graphs?

Students should write their conclusions. Ask them to reflect on their findings and compare them with their predictions. Allow time in class for students to share their findings and conclusions.

After students have compared the graphs for their nine test flights, ask them to think of a way to display their data to show which load position was the overall distance winner. You may want to discuss finding median and mean distances for each load position, if appropriate.

Students can use calculators to find the total, median, or mean distance to be used to determine the overall distance winner. Once they have an appropriate measurement for each load position, students should construct a bar graph to display their findings.

Class Conversation

- What differences do you notice about the two graphs? Are there any similarities?
- Which graph shows more information?
- Where were the paper clips when your airplane traveled the farthest?
- What conclusions can you draw about the location of weight in relation to flight distance?
- Were your predictions close to the actual results? Explain.

FURTHER EXPLORATIONS

- Students can repeat their test flights for front, middle, and back loads using a stopwatch to measure the number of seconds their airplanes remain in the air. They can use these data for their analysis.
- Have students explore other ways of distributing the paper clips on the airplane and the related effects on flight distances. They can collect new data and make new graphs or extend their existing multiple-bar graph.

School-Home Connection

Encourage students to share the activity with their families and collect data from further explorations at home. Students can make take-home tape measures using string or twine with masking-tape labels to mark equal intervals. Some families may prefer to measure the airplane distances using customary units of measure.

RELATED ACTIVITIES IN MISSION MATHEMATICS

Bar graphs are also included in "Covering an Airplane." Concepts of median and mean are developed in "Long-Distance Airplanes" and "Target Practice."

NCTM Assessment Standards: *Each team or flight crew can report its findings and conclusions orally to the class, which gives students an opportunity to plan and rehearse a group presentation. Allow students to enhance their understanding of using graphs as a tool for communicating by reporting their findings. Make the assessment an open process by having students help create a checklist as they prepare and rehearse their presentation.*

DATA-COLLECTION SHEET

My Data

Flight Number	Front-Load Distance (meters)	Middle-Load Distance (meters)	Back-Load Distance (meters)
1			
2			
3			
Total			

TRIPLE-BAR GRAPH

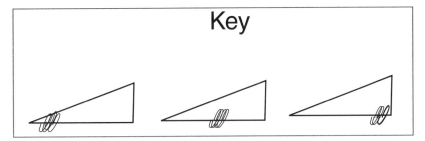

Airplane Flight Distance

Distance (in meters)

Flight 1   Flight 2   Flight 3

Conclusions _____
_____
_____
_____

Location of Load

FLIGHT DIRECTION

K 1 2 3 4 5 6

NCTM Mathematics Standards

- *Statistics*
- *Fractions and Decimals*
- *Number Sense and Numeration*
- *Patterns and Relationships*
- *Measurement*

Materials: *Paper, resource pages 32 and 33, masking tape, crayons*

Management Tip: *This activity is intentionally planned to take several periods. Determine the number and duration of sessions appropriate for your class.*

SR-71 Blackbird

NASA connections can be found in the introduction for the "Aeronautics" section.

Students experiment with the position of movable ailerons on a paper airplane to discover the effects on flight. As they perform a series of test flights for each change in the ailerons, students observe and record landing locations in a designated target area. Students display their test-flight data in circle graphs and interpret the graphs to make conjectures.

PURPOSES

- To measure and mark grids for targets
- To make and test predictions about the effects of ailerons on flight direction
- To record landing locations on a grid
- To collect, organize, display, and interpret data
- To express test-flight results as fractions and decimals
- To construct a circle graph
- To look for patterns in data to make conjectures
- To suggest future experiments to test conjectures

INTRODUCTION

NASA scientists and pilots constantly experiment with test designs and materials to make the best airplanes. They work to improve safety, to increase performance, and to reduce costs. Their experiments are done over long periods of time. They change their experiments by varying the conditions.

In this activity, your students conduct an extended experiment in which they change and test different flight conditions. By working in pairs or small groups, they will better understand how research teams of NASA scientists must work together cooperatively to complete large projects.

BEFORE THE ACTIVITY

Make overhead transparencies of the airplane diagram and resource pages 32 and 33: Data-Recording and Circle-Graph sheets 1 and 2.

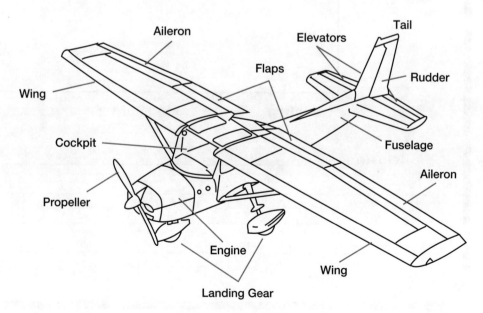

Help students mark off several target areas on the floor. The target areas should be made with two pieces of masking tape placed at right angles as shown. Each piece of tape should be at least 4 feet long. As students lay out each target area, be sure they allow enough space to stand at least 10 feet away from the target to do their test flights.

GETTING STARTED

Show your class the diagram of a propeller-driven airplane on the overhead projector and discuss different parts of the airplane: propeller, engine, fuselage, cockpit, landing gear, wings with ailerons and flaps, and tail with rudder and elevators. Present the following background information.

Airplanes are equipped with special control surfaces to give the pilot a way to change the direction and altitude of flight and to slow the airplane for approach and landing.

To raise or lower the nose of the aircraft in flight, the elevator control surface must be changed. The up or down movement of the nose of the aircraft is the *pitch*.

Turns are aided by moving the ailerons. One aileron is raised while the other is lowered. If the left aileron is down, the right one is up. If the right aileron is down, the left one is up.

To move the nose of the aircraft from side to side, the rudder on the vertical stabilizer of the tail is moved from side to side.

To have a smooth change of direction for the airplane, the rudder and the ailerons are used at the same time. The nose's movement from side to side is the *yaw* and the wing tips' up-and-down position is the *bank*.

Flaps are movable sections on the wings, which the pilot uses to fly more slowly for take-off and landing.

Class Conversation

- If you were a pilot, in what different directions would you want your plane to move?
- What movable parts do you think the pilot can control from the cockpit?
- Do you think that any of the parts shown in this picture control the direction in which a plane can travel?
- Can we do anything to a paper airplane to control its direction?

DEVELOPING THE ACTIVITY

Part 1: The Control Flight

Distribute paper and ask each student to make an airplane similar to the one shown.

Demonstrate how to fly the airplane into the target and how to record an X on the appropriate grid of data-recording and circle-graph sheet 1 to show the location of each landing.

Before students cut ailerons in the wings, ask them to work together in pairs to conduct a series of ten test flights of each partner's airplane. The series of ten test flights is important for recording results as decimal numbers. One student flies his or her airplane into the target spot from 10 feet away, and the other student records each landing. Then they reverse roles. After all students have completed this first series of test flights, bring the class together to discuss their

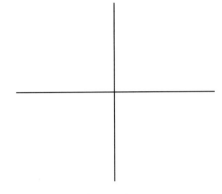

Target Areas

NCTM Teaching Standards:
Teachers must decide the depth of content used in a discussion. To simplify the activity, use the word flaps *instead of* ailerons. *However, if appropriate for your class, differentiate the roles of flaps and ailerons.*

NCTM Assessment Standards:
Observe which students can use a data-collection form as directed and which students need to learn how.

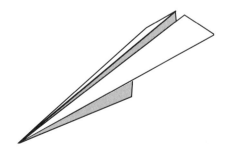

Flight Direction

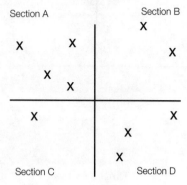

Sample Data

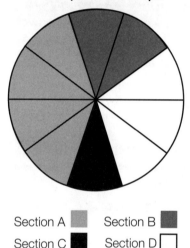

Sample Circle Graph

Section A ▨ Section B ▨
Section C ■ Section D □

NCTM Assessment Standards:
Talking and writing about solutions help students confirm their learning. To close part 1, ask students to summarize the processes of expressing their data as fractions and decimals as well as of displaying their data in circle graphs.

You can informally assess students' understanding of controlling variables in an experiment by asking why it is important to collect data about where their airplane lands before they change the position of the ailerons.

findings, to learn how to summarize their results as fractions or decimals, and to learn how to display their data in a circle graph.

Class Conversation

- Which section of the target did your plane hit the most? The least?
- What patterns do you notice in the data for your airplane?
- Of your ten landings, how many were in section A? In section B? In section C? In section D?
- What fraction can you write to express the number of landings in section A? In section B? In section C? In section D?
- What decimal can you write for each of these fractions?

Talk about how the number of landings in a section can be expressed with either a fraction or a decimal. Demonstrate how to write fractions and decimals for the given sample data. Write an equation to show the equivalence of each fraction and its corresponding decimal.

You may want to organize this part of the lesson in a table.

Table of Sample Data for Test Flights

Area	Number of Landings	Total Flights	Fraction	Decimal	Equation
Section A	4	10	$\frac{4}{10}$	0.4	$\frac{4}{10} = 0.4$
Section B	2	10	$\frac{2}{10}$	0.2	$\frac{2}{10} = 0.2$
Section C	1	10	$\frac{1}{10}$	0.1	$\frac{1}{10} = 0.1$
Section D	3	10	$\frac{3}{10}$	0.3	$\frac{3}{10} = 0.3$

Next, introduce or review how data can be displayed in circle graphs to show parts of a whole. Demonstrate with the sample data how to color and label a circle graph for the fractions or decimals that summarize the landing results for each section of the target.

Conclude part 1 of the activity by having students write the fractions and decimals that summarize their landing results for the ten test flights without ailerons. Each student should color and label a circle graph for his or her data. Encourage students to share and compare their graphs.

If you plan to do part 2 at a later time, ask students to save their airplanes and data-recording and circle-graph sheet 1 for the next session.

Part 2: Flights with Changes

Ask students to recall the names of the parts of an airplane. Also ask them to recall how to measure or estimate a 1-inch length.

Show students a sample airplane in which ailerons have been cut as shown. Explain that they are going to experiment by changing the position of ailerons to see how the change affects the direction of the airplanes. Then present this task.

Notice where the ailerons are on the wings of this airplane. Make an aileron on each wing of your airplane by making cuts in about the same locations. On each wing, the two cuts should be about 1 inch long and should be about 1 inch apart.

Management Tip: *Plan different sessions for each series of test flights in part 2.*

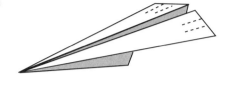

Class Conversation

- What effect do you think the ailerons will have on the landing patterns?
- Do you think that your data will show any different patterns with the ailerons in different positions? Why?
- What predictions would you like to make?

Use both data-recording and circle-graph sheets as a guide for each series of ten test flights that students conduct with the ailerons in different positions. Students should follow the same procedures as in part 1 when conducting the series of test flights for each airplane. For each different position of the ailerons, partners should display the landing results in a separate circle graph.

CLOSING THE ACTIVITY

When all data have been collected, help students organize and analyze their findings. They can cut out their circle graphs, put all graphs with the same conditions together for the whole class, look for patterns, sort the graphs, and glue them onto a class chart. They can label the categories on the chart and write their findings and conclusions.

Class Conversation

- How can we organize our graphs to compare our findings?
- What do you see in your graphs?
- Are there any similarities or differences among the graphs? Which ones? Why do you think so?
- Using your graphs, what conjectures can you make about how ailerons affected the direction of our planes? How do your conjectures match your predictions?
- What do you think happens to the air as it hits the ailerons? How do pilots use ailerons on real airplanes? Why?

FURTHER EXPLORATIONS

- Ask students what other experiments they would like to conduct to further test their conjectures about flight direction.
- Students may want to design and test different airplanes by varying the width of the wings and the size and placement of the ailerons.

RELATED ACTIVITIES IN MISSION MATHEMATICS

Different graphs are used in the other activities of the "Aeronautics" section, including line plots, stem-and-leaf plots, picture graphs, pictographs, bar graphs, and histograms.

DATA-RECORDING AND CIRCLE-GRAPH SHEET 1

Flight Results: No Ailerons

Section A | Section B

Section C | Section D

Circle Graph: No Ailerons

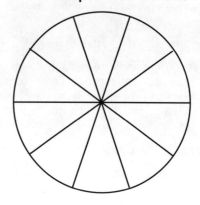

Section A—Red   Section B—Blue
Section C—Green Section D—Yellow

**Flight Results:
Left Aileron Up and Right Aileron Down**

Section A | Section B

Section C | Section D

**Circle Graph:
Left Aileron Up and Right Aileron Down**

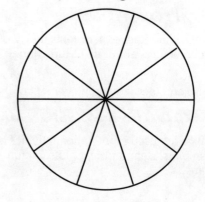

Section A—Red   Section B—Blue
Section C—Green Section D—Yellow

DATA-RECORDING AND CIRCLE-GRAPH SHEET 2

Flight Results:
Right Aileron Up and Left Aileron Down

Section A   Section B

Section C   Section D

Circle Graph:
Right Aileron Up and Left Aileron Down

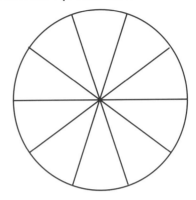

Section A—Red   Section B—Blue
Section C—Green   Section D—Yellow

Conclusions _____

_____

_____

_____

_____

_____

_____

_____

_____

_____

Flight Direction

RESCUE-MISSION GAME   K 1 2 3 4 5 6

Students play a game to learn about the four forces of flight: lift, drag, thrust, and gravity. Before playing the game, students conduct a probability experiment with spinners and record their results in tally tables and bar graphs. They use their findings to select spinners with the greatest probability of helping them win the game. Students use ordered pairs to plot points on a coordinate grid to show their flight path.

NCTM Mathematics Standards
- *Geometry and Spatial Sense*
- *Statistics and Probability*
- *Number Sense*
- *Patterns and Relationships*

Materials: *Resource pages 38–40, crayons, paper clips, graph paper, pencil*

UH-60 Black Hawk helicopter

PURPOSES

- To identify and use the four forces of flight
- To collect, organize, and interpret data
- To construct tally table and bar graphs
- To determine the likeliness or probability of success
- To read and write ordered pairs
- To use ordered pairs to plot points on a grid

INTRODUCTION

When we look at large airliners and helicopters, it seems impossible for such huge objects to lift off the ground and fly. Flight is possible because of four forces (pushes or pulls) that act on the aircraft.

Two of the forces are lift and gravity. Lift is the upward force that works against the force of gravity, the force that holds the aircraft down. Lift is created by the effect of airflow over and under the wings of airplanes or the blades of helicopters. Wings are usually thicker on the front edge and thinner on the back edge. This shape allows the air moving over the wing to move faster and, consequently, to have less pressure. The air moving under the wind travels more slowly and results in more pressure pushing up on the wing. Thus, the force pushing up on the wing is greater than the force pushing down.

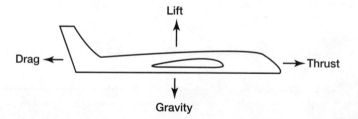

The other two forces of flight are thrust and drag. Thrust is the push or the pull forward that causes aircraft to move. Thrust is created by the engines. Drag is an opposite force that slows the aircraft. Drag is caused by the surfaces of the aircraft that interrupt or deflect the smooth airflow around the aircraft. Some things that affect the amount of drag are the flaps; the ailerons; and the size, shape, and position of the wings.

GETTING STARTED

As you introduce or review the forces of flight, ask questions to focus students' attention on a diagram with arrows to show the direction associated with each of the four forces. For example, ask which force pulls things to the ground.

As you show resource page 40, tell the class about the Rescue Mission game they will be playing. They are pilots of rescue helicopters. Their mission is to fly their helicopters to the top of a mountain to rescue lost hikers.

Show resource pages 38 and 39 and explain how the spinner determines in which direction to move. For example, if the pointer lands on Lift, students move their helicopter up one space. Ask them in which direction they should move if it lands on Drag. [Left] On Thrust? [Right] On Gravity? [Down]

When students cannot move in the direction indicated by the spinner, they stay in the same position for that turn. Show students the starting point of the game and ask them to think about the flight path for their mission. Ask them in which directions they must go to reach the mountaintop. [Up and to the right] Ask which forces will be most helpful. [Lift and thrust] Why?

Management Tip: **Students can use a pencil and a paper clip as their "propeller" at the centers of the spinners.**

DEVELOPING THE ACTIVITY
Part 1: Getting Ready for the Mission

As you distribute copies of resource pages 38 and 39, the spinners, explain that since the lost hikers are cold and hungry, the pilot needs to get to the top of the mountain quickly. Ask students to compare the spinners.

Class Conversation

- How are they alike? How are they different?
- Which spinner do you think will help your helicopter get to the mountaintop the fastest? Why do you think so?
- How can we test the spinners to check our predictions?

After discussing their suggestions, ask students to work with a partner to spin each spinner fifty times and to record all results in a tally table. When all data have been collected, help students display their data in bar graphs.

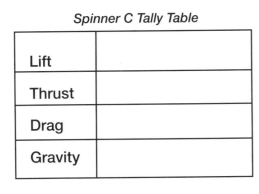

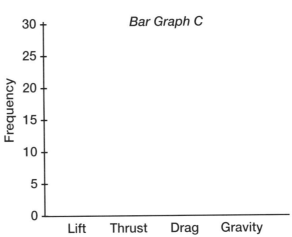

Each bar graph should be discussed and interpreted. Help students see that on spinner C, all forces have the same chance for the pointer to land on them. Since there is 1 chance out of 4 equal chances that the pointer will land on Lift, the probability is 1/4.

A sample conversation for spinner C follows.

Class Conversation

- Compare the regions in spinner C. How many of the same size do you see? [4]
- How many different forces are on spinner C? [4]
- Is it less likely, equally likely, or more likely that the pointer will land on Lift than on Gravity, Drag, or Thrust?
- What is the likelihood, or probability, of landing on Thrust? On Drag? On Gravity?

Rescue-Mission Game

NCTM Assessment Standards:
Students can write their conclusions and support them with the findings from their analysis and interpretations of the data they have collected and displayed.

- Does the pointer have the same chance of landing on each force? Why do you think that?

Next, ask students to look closely at their graphs for spinner C to interpret the results of their experiment.

- Since the probability is the same for each force on spinner C, what should the graphs look like? Why?
- Do your bar graphs show this?
- Which bars are taller? Shorter?
- What does the bars' appearance show you?

Repeat this analysis process with each spinner.

- When it is your turn, how can you use your graphs to help you decide which spinner to use on that turn?

Students should keep all spinners for the Rescue Mission game.

Part 2: Playing the Rescue-Mission Game

Introduce or review how to read and write ordered pairs to name a location on a coordinate grid. Practice using the spinners to determine moves on the game board. Ask students to predict how many spins will be needed to reach the top of the mountain.

Students then play the game with a partner to see which helicopter can rescue the lost hikers on the mountaintop first. As they take turns, students should record the following data:

- Which spinner is selected for the turn
- Where the pointer lands (lift, thrust, gravity, or drag)
- How the student moves (up, right, down, or left)
- The ordered pair that names the point to which they move

Sample Recording Sheet

Turn Number	I Selected This Spinner	Pointer Landed On	Direction in Which I Moved	Place Where I Landed
1	C	Lift	Up	(0, 1)
2	D	Thrust	Right	(1, 1)
3	D	Lift	Up	(1, 2)

They can record the flight path on the game-board grid by plotting the points for each student in different colors. When finished, students can connect the points to show the flight path.

CLOSING THE ACTIVITY
Class Conversation

- Did it take more spins, fewer spins, or the predicted number of spins to reach the mountaintop? How can you explain this result?
- Look at the ordered pairs of numbers that name the points where you landed. Do you notice any patterns in these ordered pairs?

- How did each partner record his or her flight paths? Explain how this graph shows what happened as you took turns.

FURTHER EXPLORATIONS

- Make a new game board that favors other spinners. Let students examine their findings for each spinner and decide which spinner will help them reach the new goal fastest. Then have them play the game to test their predictions.

- Make up a new spinner. Have students make a new game board that would best suit this spinner. Have them play a game to test their hypothesis. For students familiar with negative and positive integers, the new game boards can extend into the other quadrants of the coordinate plane.

- Present this problem: Suppose that a helicopter uses three gallons of fuel for each turn needed to rescue the hikers. Calculate and graph the amounts of fuel each partner used in each game played.

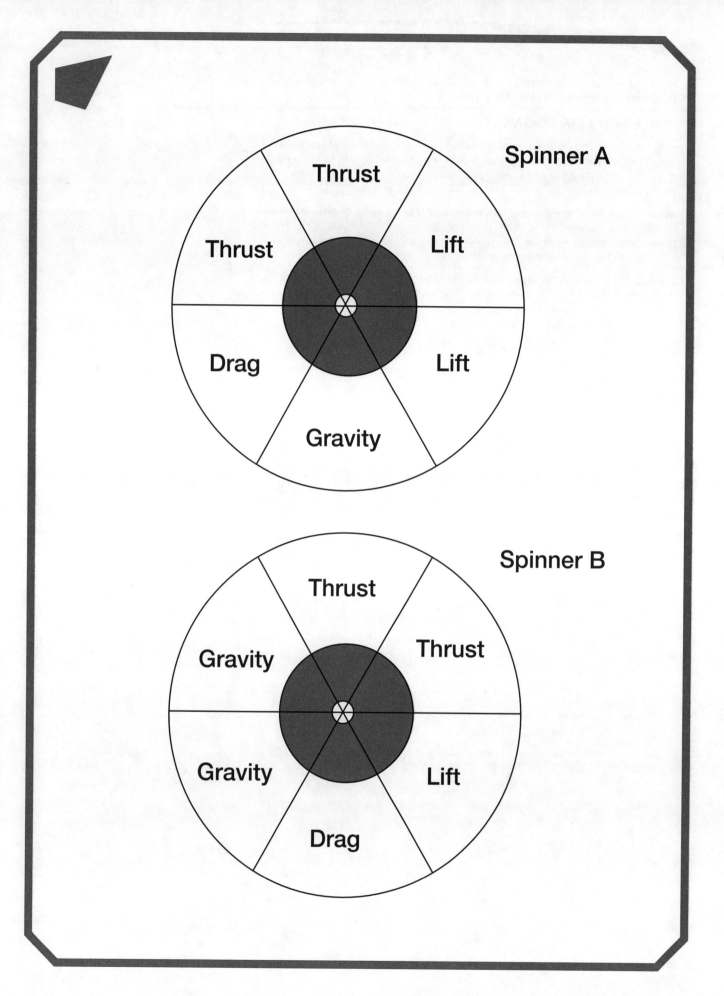

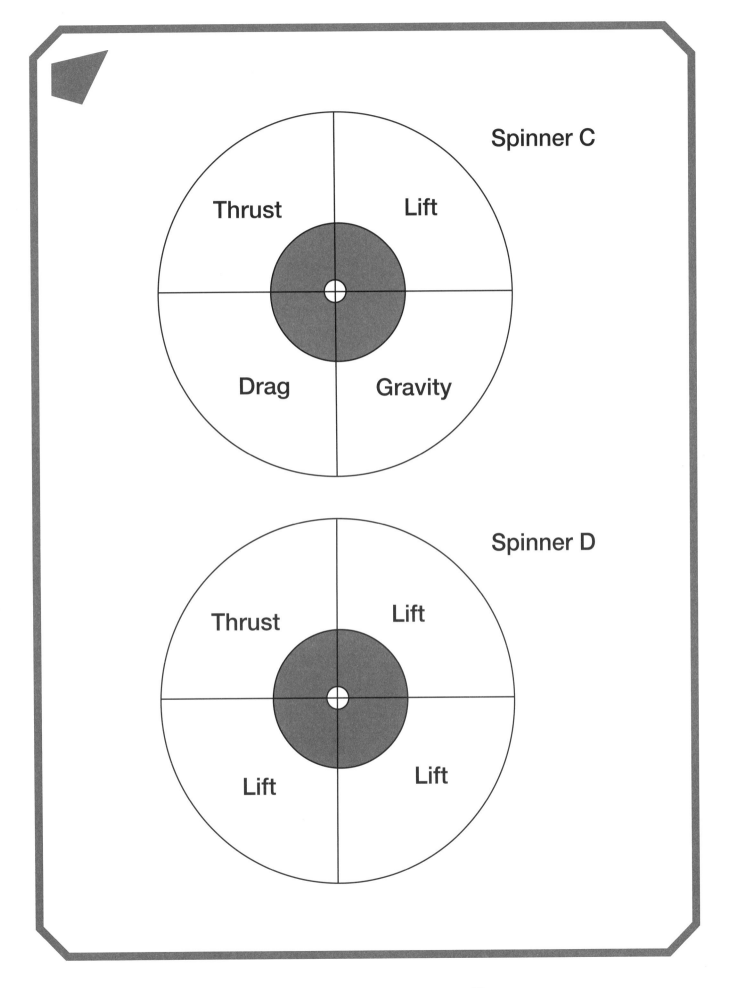

Rescue-Mission Game

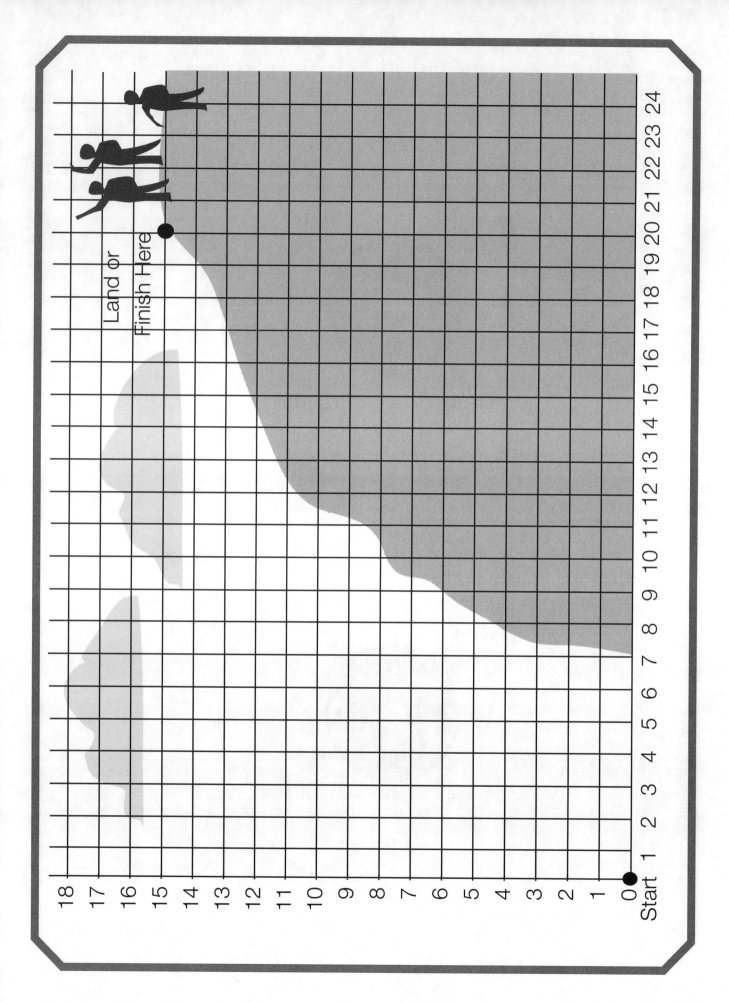

THE AIR SHOW   K 1 2 3 4 5 6

Students construct an airplane that can succeed in each of these categories: distance, air time, accuracy, and acrobatics. They fly their airplanes and record their results on a data-collection sheet. Each student displays the data for his or her airplane in a glyph by decorating a drawing of an airplane with symbols, patterns, and colors, as agreed on by the class.

PURPOSES

- To design a paper airplane that can perform well in several events
- To measure distance and time
- To calculate median scores
- To collect and organize data
- To construct and interpret a glyph

INTRODUCTION

"The Air Show" is a culminating activity for students to review and practice many mathematical concepts. It can also be an excellent opportunity to remind your class that like the pioneers of aeronautics, the men and women who work today in the aeronautics centers of NASA are always experimenting to discover new ways to make better airplanes.

Before a successful Space Shuttle could be built, NASA contributed to the development of several generations of experimental research aircraft. Show pictures of a few of NASA's experimental airplanes:

- Bell Aircraft Company's X-1, which in 1947 was piloted by Major Charles "Chuck" Yeager of the United States Air Force and became the first aircraft to break the sound barrier.
- The North American X-15 was tested in the mid-1960s. It reached a top speed of 4 534 miles per hour and a maximum altitude of 354 200 feet.
- Lockheed's SR-71 Blackbird, developed in the late 1960s and early 1970s, can travel at more than three times the speed of sound and at an altitude of 80 000 feet. It is used as a reconnaissance airplane by the United States Air Force.

GETTING STARTED

Divide the class into teams to do the following tasks. Have students design and prepare a data-collection sheet. They should plan for three test flights in each of the four events. Next, show them how to set up the flight paths or runways for the distance event. They can measure and mark distances with chalk or tape. Students then set up a target area for the accuracy event with different point values for each section of the target. Tell students that if an airplane does not hit the target at all, they should write 0 on the data-collection sheet for that test flight. Finally, help students decide on the rules for scoring the events and make ribbons for the winners. Many measurement and writing skills can be reviewed and reinforced as they communicate the rules and make awards or ribbons. Computer software may be available in your school for students' use in making certificates.

DEVELOPING THE ACTIVITY

Each student should design and make his or her own airplane that will perform well in all four events. On the day of the big event, ask students to work in pairs to collect data from three test flights in each of these events:

NCTM Mathematics Standards
- *Statistics*
- *Measurement*
- *Patterns and Relationships*
- *Number Sense*

Materials: *Paper, stopwatch or clock with a second hand, measuring tape, crayons or markers*

X-1

X-15

SR-71 Blackbird

- Air time: How many seconds will the airplane stay in the air?
- Distance: How far will the airplane travel?
- Acrobatics: How many flips or rolls will the airplane do?
- Accuracy: Can the airplane land in the target to score points?

Students find the median scores for each event by crossing out the high and low scores of the three test flights and circling the median, or middle, score.

Displaying the Data with Glyphs

After the events are completed and the scores are recorded, introduce or review glyphs, which are just a different way to show several types of data at one time. Students' glyphs will show how well their airplanes performed in all the events of the air show.

Class Conversation

Share a sample glyph and key with the students and ask questions to help them understand how each event is represented.

- What shape is the eye on the airplane? What does that represent?
- What does the mouth tell us?
- What event does this pattern of stripes describe? How well did this airplane do in the accuracy event?
- What color is the airplane? What does this tell us?
- Should all the glyphs look the same in our class, or will each one look different? Why?

Then ask the class to work together to create a key for all students to use on their glyphs.

Management Tip: *If glyphs are a new concept for your class, let students work together in pairs or small groups to help one another.*

Sample Glyph Key			
Eye	**Mouth**	**Stripes**	**Color or Pattern**
Median Distance (in feet)	**Median Air Time (in seconds)**	**Median Target Accuracy Score**	**Number of Flips**
0 to 10	1 or 2	0	0 red
11 to 20	3 or 4	1	1 green
21 to 30	5 or 6	2	2 blue
31 or more	7 or more	3	3 or more purple

Using the given key, help the students understand that this glyph shows that the airplane traveled less than 11 feet, stayed in the air for 3 or 4 seconds, is not very accurate because it missed the target, and did two flips or rolls.

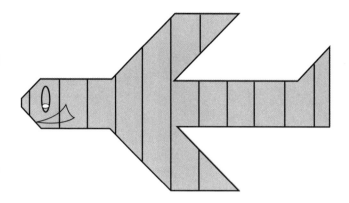

CLOSING THE ACTIVITY

When all glyphs are complete, allow students to share their glyphs with the class. Each student can present his or her glyph and challenge the class to interpret it. Then ask students how their glyphs are similar or alike. How are they different? What patterns do the glyphs help them see in their data?

ASSESSING THE ACTIVITY

This activity affords opportunities for different types of assessment. Some suggestions follow:

- Help students make simple checklists of what to include in their presentations as they share their glyph with the class.

- Encourage students to self-evaluate such things as their team's assigned task, their work with a partner to collect data, and their understanding of how to make a glyph with a set of data.

- If appropriate, ask students to write a paragraph describing what their glyph shows. They could also write a paragraph about their partner's glyph.

FURTHER EXPLORATIONS

- Students can put all the glyphs together and then make up ordering or sorting activities for their classmates.

    Example 1: Line up all airplanes in descending order for air time. Put the airplanes that stayed in the air the longest at the right.

    Example 2: Find all airplanes that stayed in the air for more than 4 seconds.

    Example 3: Find all airplanes that were the most accurate and that flew the greatest distances.

    Example 4: Find all airplanes that were the most accurate or that flew the greatest distances.

- Ask students to think of other types of data they would like to use to make glyphs. For example, students can make glyphs about themselves that might include such data as age, height, number of siblings, and number of pets.

RELATED ACTIVITIES IN MISSION MATHEMATICS

Paper-airplane flights are used to collect and display data in "Long-Distance Airplanes," "Target Practice," "Flight Direction," and "Location of Load."

NCTM Teaching Standards:

Teachers pose questions and problems that both elicit and challenge students' thinking. Examples 3 and 4 are more advanced tasks and can be used to challenge students to use mathematical reasoning.

Help students learn about using the words and *and* or *in compound sentences. Students can use chalk or string to make large circles for a Venn diagram to represent sets of airplanes that intersect and sets that are disjoint, that is, do not intersect.*

The Air Show

HUMAN EXPLORATION AND DEVELOPMENT OF SPACE

One of the activities that characterize the human race as a species is the desire to explore the world. At one time, oceans were a formidable barrier to exploring the planet; humans confined their curiosity to the land mass around them. Brave explorers ventured out onto the seas to search beyond the horizon when the level of curiosity and the promise of increased riches reached a critical point. The results of those journeys validated the cost and the risk. Through exploration, travelers learned about new cultures and new animals and plants as well as the mineral riches of new lands.

We live in an era in which the new environment to be explored is space. Our motivations to assume the cost and risk of manned exploration of space are very much like those of our ancestors. We want to learn how to survive in the environment of space, we want to identify and assess opportunities for the commercial benefits of human space flight, and we want to expand our understanding of Earth and of our neighbors in the solar system.

NASA has accepted the challenge of the human exploration of space. To this purpose, NASA strives to enable routine operating capability within the inner solar system, to explore this area of space regularly, and to support the establishment of permanent, self-sufficient settlements in space. To reach these long-term goals, NASA has methodically undertaken activities that build our understanding of exploring, using, and enabling the development of space.

Through exploring space, NASA aims to learn how to travel to a destination and, in the case of human exploration, to return safely to Earth. Safe and reliable access to orbital flight by humans has become somewhat routine through the Space Shuttle program. Groups of pilots and scientists called astronauts routinely ride the Shuttle into orbit to study the effects of space travel on humans and the advantages of performing certain laboratory and commercial activities in microgravity. Astronauts have ventured to the Russian space station, *Mir,* and Apollo astronauts have set foot on the Moon. These events are important benchmarks for the establishment of the International Space Station, where scientists will be able to live and carry out research on a long-term basis. This research may enrich life on Earth through the efforts of men and women living and working in space.

Robotic reconnaissance missions to destinations in space are important, too. These advance visits by robots help NASA characterize and map a destination. This endeavor will ensure that data are available on the composition of the Moon, Mars, and asteroids before humans begin purposeful, perhaps manned, activities on these celestial bodies. Specifically, in 1966, *Surveyor 1* was the first NASA soft-lander on the surface of the Moon. It provided important information to pave the way for the Apollo manned exploration of the Moon later in the decade. The next lunar mission is the *Lunar Prospector,* which will be inserted into low orbit over the Moon's surface to map that body and gather data about its composition. *Galileo,* launched from the Space Shuttle in 1989, returned the first resolved images of asteroids while in transit to Jupiter, and ten Mariner missions have furnished data to help NASA prepare for a series of upcoming Mars Surveyor missions. Data from these missions have increased and will continue to increase the safety of space travel and have indicated potential ways in which travel to these destinations may have commercial value.

AN INTRODUCTION TO HUMAN EXPLORATION AND DEVELOPMENT OF SPACE (HEDS) ACTIVITIES

All activities in this section integrate the four NCTM process standards: problem solving, mathematical reasoning, mathematical communication, and mathematical connections. Each activity also incorporates several of the NCTM mathematics standards.

The following brief outline describes the "HEDS" activities and the associated mathematics.

Activity Name	Mathematics	Grade Levels
Fizzy-Tablet Rockets	Fractions, mixed numbers, decimals, measuring	K–6
Scrumptious Veggie Shuttle	Ordinal numbers, sequencing, patterns	K–6
Tiles and Tessellations	Congruent figures, translations, reflections, rotations	2–6
Water, Water	Mass, capacity, volume, estimating	3–6
Living Areas	Estimating, spatial and proportional reasoning	3–6
Spheres in Space	Circles, spheres, estimating, measuring diameters, circumferences, nets	2–6
Destination: Space Station	Ordered pairs, coordinate grid	K–6

These activities can help students learn about HEDS projects at NASA. The goals are to increase our knowledge of nature's processes using the environment of space, to explore and settle the solar system, to achieve routine space travel, and to enrich life on Earth by enabling people to live and work in space.

The advanced development of NASA's Space Shuttle program has enabled us to participate in life and work on the Russian space station, *Mir*. This spacecraft is commonly known as International Space Station Phase 1. It is a test site for international cooperation in space, long-term data collection in biological and materials studies, hardware, construction methods, and the operation of many transport vehicles.

Space Shuttle Atlantis *docking with Russian space station,* Mir

The goal for the International Space Station is to have elements from the United States, Europe, Canada, Japan, and Russia. The station will include a permanent laboratory for scientific experiments involving thirteen nations: the United States, Canada, Italy, Belgium, Netherlands, Denmark, Norway, France, Spain, Germany, the United Kingdom, Japan, and Russia. Some areas of research include medical applications, the development of lighter and stronger superalloys, energy conservation, communications, computer software, heating and cooling systems, safer chemical storage and transfer processes, air- and water-purification management, and the growth of protein crystals.

Other sources of information on space exploration and the related technologies that have made it possible are available through NASA's home page on the Internet. For more information about NASA's materials for teachers, see appendix A.

FIZZY-TABLET ROCKETS   K 1 2 3 4 5 6

Students launch film-canister rockets to investigate the effects of varying the amount of fuel (fizzing antacid tablets). They observe and collect data to see whether there is a difference in time from fuel ignition to landing. Three different amounts of fuel are tested: one whole tablet, one-half tablet, and one-fourth tablet. The amount of water in the canister remains constant at one tablespoon.

Two versions of the activity are given, one for young students and one for students with more mathematical experience. Select those tasks and problems that are appropriate for your class.

PURPOSES

For all students:

- To develop number sense for, and concepts of, fractions, mixed numbers, and decimals
- To use two standard units of measure, tablespoons and seconds
- To collect, organize, analyze, and interpret data

For intermediate students:

- To use rounding to relate decimal values to whole numbers
- To find the range, median, and mean of a set of data

INTRODUCTION

Military rockets have a very long history in China, the Middle East, and Europe. They came to the United States in the War of 1812. There were some attempts to use rockets in the Civil War in the 1860s. The first flight of a liquid-fueled rocket was in March 1926, when Robert Goddard launched a rocket from his Aunt Effie Goddard's farm in Auburn, Massachusetts. The flight went the same distance as the Wright brothers' first manned flight, about 152 feet.

Robert Goddard and Wernher von Braun are the most famous pioneers of modern rockets. Their work built the foundation for the development of such well-known rockets as the V2, the Redstone, the Jupiter, the Saturn, and the Space Transportation System (STS), now commonly known as the Space Shuttle.

BEFORE THE ACTIVITY

Ask a local film processor to save clear plastic film canisters, not the black ones, which most processors are happy to do. The lids on the clear canisters fit inside the rim rather than over the side. This makes a tighter seal and allows more time for the pressure to build up inside the canister, thus producing a better rocket.

Precut the antacid tablets into halves and quarters by scoring the tablet with a serrated knife and breaking it on the line.

Primary-Grades Version (K–2)

The focus of the activity should be on fractional parts of the whole: one-fourth tablet, one-half tablet, and one whole tablet.

GETTING STARTED

As you prepare the tablets, students should see the process of dividing the tablets into the fractional parts. Demonstrate how two one-half pieces combine to equal a whole. Prompt students to investigate to find other relationships by

NCTM Mathematics Standards

- *Fractions and Decimals*
- *Number Sense*
- *Patterns and Relationships*
- *Statistics and Probability*
- *Measurement*

Materials: *Clear plastic film canisters, fizzing antacid tablets, water, one-tablespoon measuring device, digital stopwatch or other timing device to measure seconds*

Management Tip: *This activity can extend over several instructional periods.*

Kennedy Space Center in Florida is the only liftoff site in the United States for human space flight. Shuttle launches are at Launch Complex 39.

asking them how many one-fourth pieces combine to equal a one-half piece and how many one-fourth pieces combine to make a whole.

DEVELOPING THE ACTIVITY

Follow these directions to launch the rocket.

Put one tablespoon of water into a film canister. One student places a one-fourth tablet into the canister, quickly closes the lid tightly, and places the canister upside down on a level surface. As the tablet is placed in the water, the student calls out to the timer, "Start!"

A different student is nearby with a stopwatch. The stopwatch is started when the tablet is dropped into the water in the canister. When the rocket hits the ground at the end of its flight, the timer stops the stopwatch.

Help students measure and record the number of seconds it takes for each flight from ignition to landing. You may want to use a large class chart to record the times for the launches with one-fourth, one-half, and whole tablets.

Be sure to do enough trials to allow students to see a pattern emerging. The one-fourth tablets take the longest, the one-half tablets fall in a middle time range, and the whole tablets are the fastest.

Safety Tip: Be sure that students waiting for liftoff stand away from the canisters after they have been placed upside down. Rockets usually fly a maximum of 3 to 4 feet. Use caution especially with the one-fourth-tablet launches. Students think that the rocket is not going to fire and want to go closer to see if anything is happening. Eventually the rocket will take off. You do not want liftoff to occur while someone is looking down on the canister!

CLOSING THE ACTIVITY

Class Conversation

- Tablets of which size take the longest time from ignition to landing?
- Tablets of which size are the fastest?
- Should we record the times for two one-half tablets and the times for whole tablets in the same column on the data sheet? Why?
- Compare the time for two one-fourth pieces and the time for one one-half piece. Are they about the same? Why?

FURTHER EXPLORATIONS

To explore the equivalency of fractions, try several launches using different combinations, such as placing two one-fourth pieces in a canister to see whether the time is similar to the times for the one-half-tablet data. Continue by comparing other equivalencies.

Intermediate-Grades Version (3–6)

The activity is the same, but the mathematics used and the tasks and problems presented are more advanced. For example, the process of finding the total number of tablets needed to do a certain number of trials should be done by the students if they have the knowledge base.

> Suppose that we want to do ten rocket launches using whole tablets, ten launches using one-half tablets, and ten launches using one-fourth tablets. How many tablets do we need for our investigation?

This rich problem is appropriate for such problem-solving strategies as *act it out* and *draw a diagram*. When finding the number of tablets needed for ten launches with one-fourth tablets, students encounter a situation for which they need more than two whole tablets but less than three whole tablets. This situation can be used to begin a discussion of divisibility.

Problem-Solving Note: Students with less experience in using problem-solving strategies may need guided instruction about the problem-solving process.

COLLECTING AND RECORDING THE DATA

Older students can be divided into teams and given more responsibility for measuring times and recording their data. If data are gathered with a digital stopwatch, have students round the flight times of each entry.

To strengthen students' understanding of how fractions are related to decimals, use the benchmarks 1/2 and 0.50. Explain that 0.50 is another name for 1/2. Lead students to understand that decimal numbers greater than 0.50 are greater than 1/2. Therefore, when a flight time has a decimal greater than 0.50, students should increase the flight time to the next greater whole number. For example, 17.93 becomes 18. Likewise, decimal numbers less than 0.50 are less than 1/2. When a flight time has a decimal less than 0.50, the decimal amount should be dropped, which leaves just the whole number. For example, 17.45 becomes 17.

ANALYZING THE DATA

Students can find the median flight time for each tablet strength. See the activity "Long-Distance Airplanes" for an easy method of finding the median of a data set. If your class does more than three launches for each tablet size, modify the given method for finding the median. First, list the times for a tablet size in order from the least number of seconds to the greatest number. Students should alternately cross off the greatest and least numbers in the list until only one number is left. With an even number of launches, students are left with two values. They should find a number midway between these values by using the *guess-and-check* problem-solving strategy.

If appropriate for your students, guide them to model and calculate the mean for each size of tablet. See the activity "Target Practice" for suggestions of how to model the mean of a data set with such manipulatives as connecting cubes.

Class Conversation

- Why is finding the mean important?

- Can you think of an example for which finding the mean for the flight times may not give you the best numbers for comparing your data? [Example: If one flight time is extremely different from the other times, it can greatly influence the mean. Sometimes it is better to compare the median, or middle, value. This measure is not as influenced by extreme scores as is the mean.]

Students usually find it very easy to determine the mode for a data set by counting to see which time occurs most frequently at each tablet size.

As students discuss different ways to compare the results of their experiment, you may want to ask them to look at how spread out or how clumped together the times are. They can find the range of the times by subtracting the shortest time from the longest time at each tablet strength. For example, if the times for the one-fourth tablets were 18, 12, and 14 seconds, the range would be 18 –12 = 6.

INTERPRETING THE DATA

Class Conversation

- Which tablet produced the longest flight?

- Why do you think the one-fourth tablet took consistently longer from ignition to landing? [The smaller tablet took longer to build up enough gas to produce the pressure necessary to blast the lid off the canister.]

- Did you observe any consistent differences in the highest point of launch, or the *apogee,* achieved by the rockets with a certain size of tablet? [Usually, no discernible differences are observed. The differences in height of launch appear to be random and more a function of extraneous variables, such as the tightness of the seal of lids or the strength of wind.]

NCTM Teaching Standards: The discourse of this lesson should encourage students to design their own data-collection sheet. This process gives students opportunities to share their understandings about the goals of the investigation. They can anticipate and discuss how to organize the data they are to collect.

Fizzy-Tablet Rockets

Assessment Notes: *After small-group or whole-class discussion, students can write a summary of the class activities. They should note that they—*

- *found the number of tablets needed to conduct the investigation;*
- *launched rockets;*
- *collected and displayed data;*
- *analyzed data to find the mean, median, mode, and range;*
- *made conclusions about their data and experiment.*

NCTM Assessment Standards: *Assessment should advance students' understanding of mathematics. Check students' conclusions to see whether they make appropriate concluding remarks about the data and the class activity. Students can include their summaries in their portfolios. If equipment and software are available, your class could add a software stack or other multimedia report of the activity to class presentations, including photographs or videotapes of the activity being conducted.*

- Which tablet size produced the fastest flight? Why?
- Which tablet size was the most consistent, that is, had about the same times?
- Which had the most different times, that is, the most variability?

FURTHER EXPLORATIONS

- For a related, but more advanced, problem, ask students this question. What if we want to do the same number of launches with whole tablets, one-half tablets, and one-fourth tablets, but we do not want to have any leftover pieces? How many launches should we plan to do in our investigation? [Sample answer: Four trials at each strength. This investigation would require four whole tablets, two tablets cut in half to make four half pieces, and one tablet cut into quarters to yield four pieces. 4 + 2 + 1 = 7 tablets.]

  Record the addition sentence and then extend the problem. How many tablets would you need to do eight trials at each strength? [8 + 4 + 2 = 14] How many tablets are needed for twelve trials at each strength? [12 + 6 + 3 = 21] What patterns do you see?

- Now that students have a sense of the range of possible times for each tablet size, try a blind launch series. In this activity the launch team decides on four tablet sizes and keeps them secret from the data team. The launch team launches its first rocket and writes the tablet size used on its data sheet as launch A. The data team measures the time from ignition to landing and records it as A on its data sheet. On the basis of the flight time, the data team decides if launch A used one-fourth, one-half, or one tablet and writes the prediction on its data sheet.

  Continue in this manner for launches B, C, and D. At the conclusion of the four launches, the launch team compares its data sheet with the data sheet of the data team to see how closely its predictions match the actual tablet sizes used. The two teams can exchange roles.

- Students may wish to conduct the investigation again using different fuel combinations: antacid tablets and vinegar or baking soda (1/4 tsp., 1/2 tsp., and 1 tsp.) and vinegar. Compare these results with results from the current investigation. Did different fuel combinations significantly alter flight times?

RELATED ACTIVITIES IN MISSION MATHEMATICS

Students learn about finding the mean of a data set in "Target Practice" and finding the median in "Long-Distance Airplanes."

SCRUMPTIOUS VEGGIE SHUTTLE

K 1 2 3 4 5 6

Students simulate a Space-Shuttle launch using an edible model.

PURPOSES

- To simulate the sequence of the flight of a Space Shuttle from launch to landing with counting, sequencing, and using ordinal numbers
- To compare the relative sizes of the Space-Shuttle assembly—external fuel tank (ET), solid rocket boosters (SRBs), and the orbiter
- To make a model of the Shuttle assembly

INTRODUCTION

Students learn about the components of the reusable Space Transportation System (STS). The Shuttle assembly has three major elements: the orbiter (length, 122 ft); the twin, white, solid rocket boosters (length, 149 ft); and the orange external tank (length, 154 ft). The orbiter and the solid rocket boosters are reused on other missions. A new external tank is needed on every flight. A Space Shuttle usually orbits about 115 to 250 miles above Earth.

The orbiters of the STS are *Enterprise,* a test vehicle built in 1977; *Columbia,* the first to fly in 1981; *Challenger,* which first flew in 1983 but was destroyed in January 1986; *Discovery,* which first flew in 1984; *Atlantis,* which first flew in 1985; and *Endeavour,* named by school children from Senatobia, Mississippi, and Tallulah Falls, Georgia, in 1989.

GETTING STARTED

Have students share what they know about the Space Shuttle by asking them the names of the Shuttles, the reasons that NASA launches Shuttles, and the location of the launch site.

DEVELOPING THE ACTIVITY

Before the activity, prepare carrots by slicing them in half lengthwise. Cut crosswise to make two large pieces with the thick end of the carrot.

This activity works better in a small group. Give students these directions for building their Space Shuttle.

- Place a carrot piece on a plate, flat side down to represent the external fuel tank of the Space Shuttle. It is the longest part of the Shuttle assembly.
- With peanut butter, attach two same-size celery sticks, one on each side of the carrot. These represent the SRBs (solid rocket boosters) of the Shuttle. They should be longer than the orbiter but shorter than the carrot.
- Use a cardboard template to cut an orbiter from white bread. The orbiter should be shorter than the celery sticks. Spread peanut butter on the orbiter and attach it to the external fuel tank (carrot).
- When all students have completed their entire assembly, it is time to simulate a launch sequence.

    Countdown: 10—9—8—7—6—5—4—3—2—1— LIFTOFF!

- Soon after launch, simulate SRB separation by detaching the two celery-stick solid rocket boosters and placing them on the plate. Next, simulate the external-fuel-tank release in a like manner.
- After orbiting Earth several times, the orbiter should land back on the plate.

NCTM Mathematics Standards
- *Number Sense and Numeration*
- *Measurement*
- *Patterns and Relationships*

Materials: *Carrots, celery, white bread, peanut butter, nonsharp knives, poster-board orbiter template, and plates. (Other foods can be substituted for the various Shuttle parts, such as bananas and bread sticks.)*

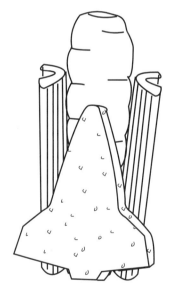

NCTM Teaching Standards: *Young students must listen to and respond to the teacher to communicate mathematical ideas.*

Note for Intermediate-Level Teachers: *This activity is excellent for older students to share with younger students. It gives them opportunities to communicate mathematics to others. Ask them to think of questions to ask the younger students to help them compare the lengths of the components of the Shuttle and understand the sequence of events in a Shuttle launch.*

The activity can also be used as a motivating introduction for an informal project related to scale drawings, scale models, and proportional reasoning. Students can use paper or cardboard cylinders to make their own models of Space Shuttles.

CLOSING THE ACTIVITY

Class Conversation

- Tell me about two parts of the Shuttle that are equal in length.
- What is the tallest part of the Space Shuttle? What is the shortest part?
- In our class, how many orbiters are there? How many fuel tanks? (Encourage counting by ones.)
- In our class, how many solid rockets boosters are there? (Encourage counting by twos.)
- What counting pattern did we use in our countdown to liftoff? Why is this a good way to get ready for the liftoff?
- In our Shuttle launch, what happened first? Second? Third? Last?

ASSESSING THE ACTIVITY

- Given a model or picture of the Space Shuttle, can students correctly label and name all the component parts?
- Can students reenact the launch sequence using a different model of the Shuttle, either commercial or student- or teacher-made?
- Ask students to draw a picture of the launch sequence for a portfolio.

FURTHER EXPLORATIONS

Use the time, event, and dialogue information found in *The Space Shuttle Operator's Manual* by Joels, Kennedy, and Larking (New York: Ballantine Books, 1988) to extend this activity for upper elementary school students. With this information, students can calculate and answer questions about time-elapse sequences in the Space-Shuttle launch event.

Space Shuttle Launch-to-Orbit Profile

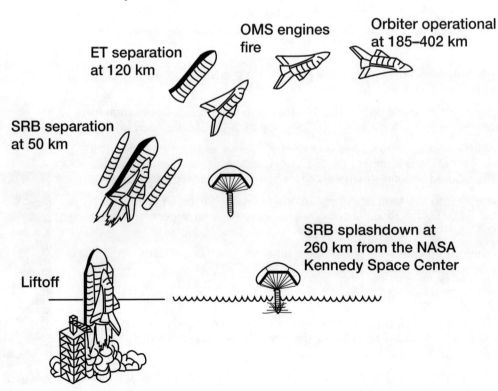

TILES AND TESSELLATIONS K 1 2 3 4 5 6

Students find combinations of geometric figures that completely cover a flat surface with no gaps or overlaps. They then apply this knowledge of tessellating, or covering, a surface to the real-life problem of tiling the underside of the Space Shuttle.

NCTM Mathematics Standards
- *Geometry and Spatial Sense*
- *Patterns and Relationships*

PURPOSES
- To identify properties of geometric figures
- To tessellate, or cover, a plane surface with congruent geometric figures
- To recognize that figures remain congruent after slides, turns, and flips

Materials: *Pattern blocks, resource page 56, a few sheets of graph paper in various sizes, quilt blocks or pictures of quilts, plastic one- or two-liter bottle with tapered top (optional)*

INTRODUCTION

As Space Shuttles return to Earth, they reenter Earth's atmosphere. During this time, the Shuttles must be able to withstand extreme temperatures from friction during reentry.

Ask students to pretend that they are NASA scientists who must design an exterior surface to protect Space Shuttles during reentry. Share with them the following essential requirements. The material must completely cover the undersurface of the Shuttle—no gaps or holes—with a smooth surface that reduces friction, insulate the Shuttle from extreme temperatures that range from –250°F while in orbit to 3000°F during reentry, and be easy to repair without the entire exterior surface of the Shuttle being replaced.

If appropriate for your students, present the following background information about the actual tiles used on the Space Shuttle.

The tiles of the Shuttles are made of silica with a ceramic coating. They are from 1 inch to 5 inches thick, but are mostly air (over 90%). Many of the tiles have six surfaces, like a brick, but some have eight to ten surfaces to cover odd-shaped places. Most tiles have top and bottom surfaces in the shape of squares that are approximately 6 inches by 6 inches. They are cut accurately to 10 000th of an inch. Each Shuttle has more than 20 000 tiles.

BEFORE THE ACTIVITY

You may want to make an overhead transparency of resource page 56. Use a copy machine or the transparency to enlarge the outline of the Shuttle to an appropriate size for your students.

If pattern blocks are not available, cardboard substitutes can be made by adult volunteers, aides, or older students. All copies of each geometric figure must be congruent.

To build an understanding of tessellations, explore such resources as the NCTM journal *Teaching Children Mathematics;* the books *Tessellations: The Geometry of Patterns* by Stanley Bezuszka, Margaret Kenney, and Linda Silvey (Sunnyvale, Calif.: Creative Publications 1977) or *Creating Escher Type Drawings,* by Ernest

R. Ranucci and Joseph L. Teeters (Sunnyvale, Calif.: Creative Publications, 1977); and software programs about pattern blocks and tessellations.

GETTING STARTED

To help students understand the concept of covering a surface with figures that tessellate, show students such patterns as floor or ceiling tiles that fit together, squares on different sizes of graph paper, quilt blocks that fit together to make a pattern, and the tessellations of artist M. C. Escher.

DEVELOPING THE ACTIVITY

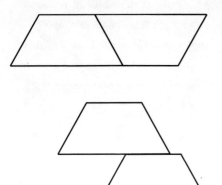

Demonstrate with pattern blocks how to tessellate a plane surface. Place blocks adjacent to one another, edge to edge. The top figure at the left shows a correct placement and the bottom figure shows an incorrect placement of blocks.

Distribute pattern blocks to groups of students and discuss the names and properties of the different pieces. Let students experiment and make conjectures to discover the best ways to cover a surface with no gaps or overlaps. Students can work in small groups to explore how they can turn, flip, and slide the pattern blocks to make them fit together to make different patterns.

Introduce or review the terms *congruent* (same size and same shape), *translate* (slide), *rotate* (turn), and *reflect* (flip), as appropriate for your students. Ask them to observe whether the figures change size or shape when they are translated, rotated, and reflected. Encourage members of the groups to talk about the properties of the different geometric figures and to help one another use proper terminology.

As students find ways that the blocks fit together with no gaps or overlaps, have them record the patterns by using dot paper or by tracing. Then pose the following problem.

> What is the best way to cover the surface of the Space Shuttle? Cover your outline of the Shuttle with pattern blocks to show your ideas.

Give each group an enlarged outline of a Shuttle on a piece of drawing paper. Ask students to use their tiling experiences to help them cover the outline of the Shuttle by tracing the shapes with no gaps or overlaps. Tell students that they may need to trace only parts of tiles at the edges of the outline to complete their tessellation.

As the class is waiting for all groups to complete their tracings, encourage those who finish to experiment with their blocks to find and record other combinations of pattern blocks that tessellate.

CLOSING THE ACTIVITY

Groups can share their conjectures, discoveries, strategies for solving the problem, and tracings with the class. Encourage students to discuss which motion or motions of the pattern blocks helped the most to cover the outline with different shapes.

As groups reflect on the activity, they can compare the strategies and solutions presented to see how they are alike and different.

Class Conversation

When all groups have completed their reports, you may want to ask the following:

- Which pattern block can be used by itself, with no other blocks of a different

NCTM Teaching Standards:
Primary-level teachers can choose which of the following vocabulary words are appropriate to review or introduce for their students: **congruent, exterior surface, hexagons, parallelograms, plane, rectangles, reflection** *or* **flip, rhombus, rotation** *or* **turn, squares, tessellate, tessellation, translation** *or* **slide, triangles,** *and* **trapezoids.**

shape, to tessellate, or cover, a surface—the triangle, square, parallelogram, trapezoid, or hexagon?

- What combinations of more than one shape can be used?
- If you had to make the tiles for the Shuttles, would it be easier to use one shape or several different shapes? Why?

ASSESSING THE ACTIVITY

Assessment opportunities include observing—

- students' use of proper vocabulary to describe properties of geometric figures;
- their willingness to experiment with slides, flips, and turns to discover tessellating patterns;
- their conclusions that hexagons, quadrilaterals, and triangles tessellate to cover a plane;
- their abilities to communicate why Shuttles must have an outer covering with no gaps or overlaps.

FURTHER EXPLORATIONS

- Using a 1- or 2-liter bottle as a model of the Shuttle, demonstrate the problems of tiling a curved surface that tapers toward the front. Explain how the shape of the bottle is similar to that of the Shuttle. Let students try to fit a smaller copy of their Shuttle tessellation over the bottle. Encourage their discussion of the need to change the size and shape of the tiles to adjust for the tapering near the nose of the Shuttle.
- Given the approximate size and number of Shuttle tiles, have students estimate and trace an area on the playground that the tiles would cover.

RELATED ACTIVITIES IN MISSION MATHEMATICS

Students may use geometric transformations such as rotations or reflections in "Covering an Airplane." However, there the emphasis is on a standard unit of measure, the square inch, and on fractions.

NCTM Assessment Standards: *Observing students is one way to verify that they are developing mathematical power, which includes the ability to make conjectures and reason logically, to solve nonroutine problems, to communicate about and with mathematics, and to connect ideas.*

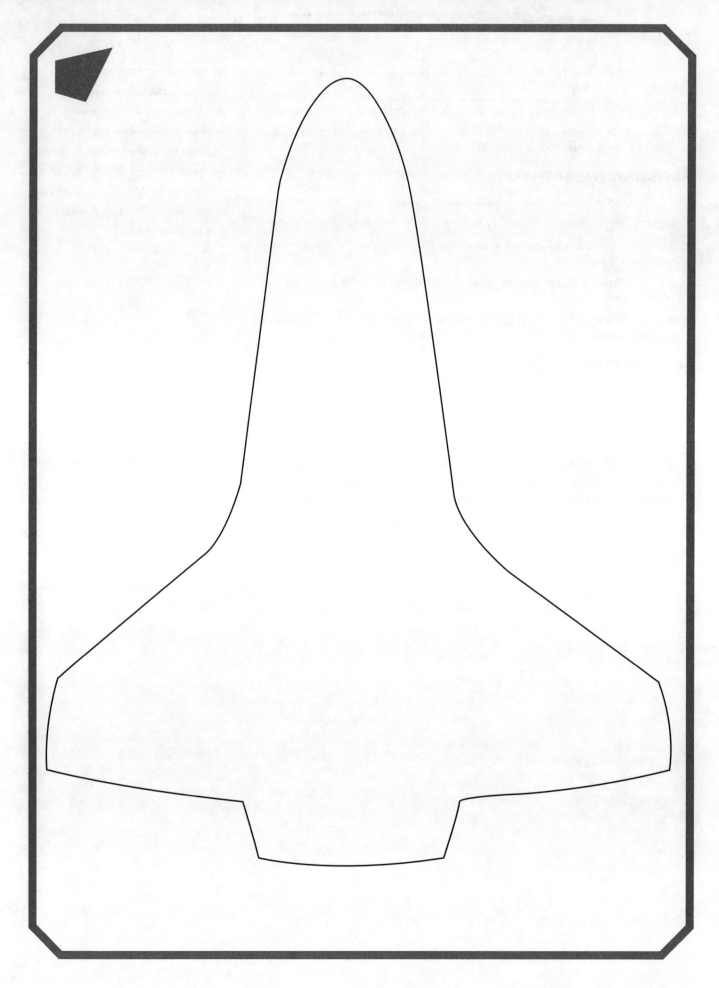

WATER, WATER

K 1 2 3 4 5 6

Students compare the amount of water they use in daily life with the amount allotted for each person each day on a Space Shuttle. They estimate and measure the heaviness of, and amount of space occupied by, a gallon of water. They collect, organize, graph, analyze, and interpret data from their investigations.

INTRODUCTION

NASA has designated that each Shuttle astronaut be allotted 6 gallons of water a day. This restriction is necessary because water is heavy. Each extra pound adds to the weight of the Shuttle at liftoff and, therefore, requires extra fuel. In addition, water takes space that can be used for other payloads and experiments.

Part 1: How Heavy Is a Gallon of Water?

PURPOSES

- To recognize a gallon as a unit of capacity
- To estimate and measure the mass of 1 gallon of water
- To determine the number of pints in 1 gallon
- To collect and organize data
- To choose an appropriate graph
- To analyze and interpret data

GETTING STARTED

Ask students to work in small groups and list all ways they use water during a day. In a whole-class discussion, compile on a chart a list from the groups. Keep the list posted.

Students should discuss with their groups how much water they use each day. Members of each group then record their estimates, and each group's estimates are reported in a new class chart.

You may want to prompt a class discussion in which students share their rationale for their estimates. Such discourse can help both you and your students assess their prior knowledge of units of measure for capacity and the relationships among those units.

Class Conversation

- Do you think that you could manage with 6 gallons of water a day?
- What changes would you have to make in the way you use water?
- How much water is eight 8-ounce glasses?
- How much do you think that you drink each day?

After students have had an opportunity to reflect on how much water they use daily, review or introduce the vocabulary words *gallon*, *pint*, and *mass*. Set a scale to zero and place a waterproof 1-gallon container on it.

DEVELOPING THE ACTIVITY

Present the following questions to the class.

- Have you ever carried a gallon of water? Did it seem heavy or light?
- About how heavy is a gallon of water?

NCTM Mathematics Standards
- *Number and Operations*
- *Measurement*
- *Statistics*
- *Estimation*
- *Geometry and Spatial Sense*
- *Patterns and Relationships*

Materials: *Large chart paper; rulers and yardsticks; cardboard; scissors; tape; waterproof containers: cup, pint, quart, and gallon; scale calibrated to the ounce; 1-gallon milk jug with the top cut off; 1-inch cubes; graph paper*

Management Tip: *These activities can be used individually or as a unit over several instructional periods.*

For health reasons, each astronaut is required to drink eight 8-ounce glasses of water a day.

- How many pints of water can we pour into a gallon container to fill it?
- How heavy is 1 pint of water?

COLLECTING THE DATA

Students take turns filling the container by pouring in pints of water, one at a time. They should stop before each pint is added and estimate the mass. After each pint is added, they should measure and record the mass and discuss their observations.

After the first pint is poured into the gallon container, ask them how heavy they think two pints will be. Have them predict the mass of three pints.

After the second pint of water is poured, ask students to evaluate how close their predictions were.

Students then add a third pint. Have them answer these questions.

- How many pints have we put in the gallon container?
- Is the gallon more than half full?
- Do you want to change your estimate of the number of pints needed to fill the gallon or your estimate of the heaviness of the gallon of water?

Students continue to fill the container until they have measured and recorded the mass of 1 gallon. When they have recorded the number of pints they used to fill the gallon container and the mass at each stage, encourage them to describe any patterns in the measurements.

GRAPHING THE DATA

Emphasize that during the investigation, students recorded several estimates and predictions. Have them name some of the things that they estimated and predicted. Encourage them to think of ways to organize and show these estimates and predictions.

Invite students to work together in small groups to solve the following problem.

> Your group is to decide how to organize and display your data from our investigation—your estimates, predictions, and measurements. Then share your graph with the class and help us interpret it.

Encourage an open discussion of different ways to show the estimates, predictions, and measures.

Class Conversation

- What are some appropriate ways to display our measures of the mass of the container as each pint of water was added?
- What do you think are the best ways to graph our estimates of the number of pints in 1 gallon? The predicted heaviness?
- What patterns or relationships can we show in our graphs?

If appropriate for your students, you may want to begin an analysis of the data by helping students find the mean of their predictions of the mass as each pint was added to the gallon container. This analysis can be done with concrete materials or calculators, depending on the prior experiences of the class in finding means.

NCTM Assessment Standards: *It is important for students to expect to assess their own progress. This assessment includes being aware of the knowledge they have or do not have to bring to a new mathematical problem or task.*

NCTM Teaching Standards: *Teachers should create a learning environment that fosters the development of each student's mathematical power by consistently expecting and encouraging students to work independently or collaboratively to make sense of mathematics.*

CLOSING THE ACTIVITY

Ask groups to share their graphs and findings with the class. Encourage class discussion of the different strategies used for displaying their data. To begin the interpretation of their graphs, ask each group what its graph shows.

The class will find that the mass of each gallon of water is about 8 pounds. Discuss with students the following factors that would give variations in the results: the care in measuring, the use of different measuring tools, and the ease and accuracy of reading calibrations on the measuring tools.

Connect the students' investigation to the astronauts' limited amount of water with the following questions:

- If 1 gallon has a mass of 8 pounds, how heavy is the water for each astronaut for one day?
- Assume that seven astronauts are in the Shuttle crew. How heavy is the water for the entire crew for one day?

Part 2: Designing Containers for Water
PURPOSES

- To estimate and measure in cubic inches (or cubic centimeters) the volume of containers that hold 1 gallon
- To compare the shapes of different containers with equivalent volumes
- To design a container to hold 1 gallon and a container to hold 6 gallons

INTRODUCTION

This activity helps students understand that water for astronauts on Shuttles does not have to be stored in containers that look like milk jugs. Other shapes of containers have the same capacity.

To help students relate to limited storage spaces, ask them to think about storing 4 to 6 gallons of water in their home refrigerator. Would this amount of water take up a lot of space in the refrigerator?

BEFORE THE ACTIVITY

Make one or two different rectangular containers with a capacity of 1 gallon, for example, with dimensions of 3" × 11" × 7" or 5.5" × 6" × 7".

GETTING STARTED

After reviewing or introducing the vocabulary words *capacity* and *volume*, show students a 1-gallon milk container. Ask them what other shapes of containers could be used to store 1 gallon of water.

DEVELOPING THE ACTIVITY

Have students find the volume of a 1-gallon jug by counting the number of 1-inch cubes (or 1-centimeter cubes) needed to fill the container. They should find that it takes about 231 cubic inches.

Next, have students explore the volume of the previously prepared containers by counting the number of 1-inch cubes needed to fill them.

Students then work in small groups to construct their own versions of containers that would have the same volume, approximately 231 cubic inches. Finally, challenge the class with this problem.

Instructional Note: *Students learn how to model the mean of a data set in the activity "Target Practice."*

NCTM Teaching Standards: *The closure for this activity uses a connection to the space program. The water requirement for an astronaut is used to pose new problems involving multiplication.*

For young students, use the approximate mass of a gallon of water. For older students, decimal notation can be used.

Instructional Note: *This activity can be set up as a learning center and be conducted over an extended time.*

NCTM Assessment Standards: *These activities can be assessed by observing and recording students' progress in the following areas:*

- *Their willingness to work together to solve problems*
- *Their use of problem-solving strategies*
- *The oral and written communication skills they use to share their ideas and strategies for solving problems*
- *Their accuracy in measuring mass and volume*
- *The correct use of vocabulary*
- *Their abilities to collect, organize, graph, analyze, and interpret data*
- *Their abilities to measure dimensions of containers, mass of containers filled with water, and volume of containers*
- *The attitudes and confidence they display when making and verifying predictions and estimates*
- *Their abilities to organize and complete a multistep project*

> Pretend that you are a group of NASA engineers planning the best uses of available space in Space Shuttles and the International Space Station. Your group is to design a container that can hold 6 gallons of water—the amount an astronaut is allowed for one day.

CLOSING THE ACTIVITY

Bring the class together. Remind them to address the problem of conserving space on the Shuttles and on the space station. Ask the groups to show their designs. Encourage each group to share its strategies and solutions for solving the space problem and to see how its design is similar to, or different from, those of the other groups.

When all groups have completed their reports, ask students how they know that a container is big enough to hold 6 gallons. About how heavy would their container be if it were filled with 6 gallons of water?

FURTHER EXPLORATIONS

- *Water-consumption investigation:* Revisit the chart created by the class to record the ways they use water, and add additional ideas. Ask the class to brainstorm ideas about water consumption that they would like to investigate. Possible suggestions follow:

  — How much liquid do we drink each day?

  — How much water is used in food preparation?

  — How much water goes down the drain while we are waiting for the warm water to come?

- *Central tendency:* Discuss the term *averages*. Students may know about the mode (most frequent), the median (middle number), and the mean (the sum of the values divided by the number of students). Talk about which *average* is most appropriate for their study.

- *Make a model:* Students record how much water they use in a day and make a physical model of the volume for this amount.

- *Conservation plan:* Students make a plan for conserving water in their daily lives or think of ways in which astronauts can conserve water in space. They pretend that they are on a Shuttle mission and record and explain their ideas in their mission log book.

- *Algebraic thinking:* Students may be able to use their models to create a formula that gives the mass of the water each astronaut would need for a mission. For example, let *n* represent the number of days. Then mass = $n \times 6$ gallons a day $\times 8$ pounds a gallon. Have students use the formula to make a table to show the mass of water for different numbers of days in the mission.

- *Water, health, and nutrition:* Help students prepare a breakfast from dried foods, such as powdered orange drink, oatmeal, cocoa with powdered milk, powdered eggs, and dried fruit like raisins or apricots. Weigh the food before and after it is prepared. Ask students to evaluate the meal by asking these questions:

  — Is this a balanced breakfast?

  — How is it related to the food pyramid?

  — What are the advantages to dried food as opposed to fresh food?

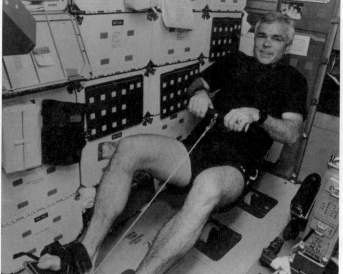

LIVING AREAS   K 1 2 3 4 5 6

Students use masking tape to outline the actual size and shape of the living area of the Shuttle orbiter on the classroom floor. In crews of five to seven members, students role-play the duties of astronauts.

PURPOSES

- To compare and relate dimensions in a diagram to actual dimensions
- To estimate distances using a standard measure, the foot
- To outline in the classroom the living area of the orbiter
- To sequence activities of a mission
- To collect, display, and interpret data from experiments

INTRODUCTION

Astronauts spend a lot of time training together to learn how to live with one another during the strenuous period when they are in space. They must cooperate to have a safe, positive environment in which to live and work.

Living in microgravity affects the human body. On Earth, the heart works against gravity to pump blood from the lower part of the body. In microgravity, where the force of gravity is greatly reduced, the face and neck appear much fuller because of extra fluids in the upper body. The waist becomes smaller and boot laces have to be tightened. Because the disks between the vertebrae no longer have the downward force of gravity on them, astronauts' heights usually increase by 2.5 to 5 centimeters in microgravity. Astronauts return to their previous height when back on Earth.

Some negative effects of living in microgravity include bone loss and the atrophy of muscles that are not exercised.

BEFORE THE ACTIVITY

You may want to make an overhead transparency for resource page 65, a floor plan of the living area of a Shuttle. In addition, the NASA video *Living in Space*, in the Liftoff to Learning Educational Videotape Series, is available from NASA CORE or a NASA ERC, listed on page 117.

Part 1: Setting Up the Shuttle
GETTING STARTED

Ask students to pretend that they are on one of the Space Shuttle missions. What do they think it would be like to live on the orbiter? How big do they think the living area of the orbiter is?

If available, show NASA's video *Living in Space*.

Discuss each area in the orbiter's interior—flight deck, mid-deck, lower life-support deck, and payload or cargo bay. Then use an overhead transparency to explain the diagram with the floor plan of the living area. Discuss the different areas and their functions.

As you use the overhead projector, show how the small image on the transparency becomes much

NCTM Mathematics Standards
- **Geometry and Spatial Sense**
- **Measurement**
- **Estimation**
- **Statistics**
- **Patterns and Relationships**

Materials: *Yardsticks, metersticks, or tape measures; masking tape; resource page 65; ribbon or string; fitted single-bed sheet (optional); cardboard boxes and newspaper (optional)*

Management Tip: *This activity can extend over several instructional periods.*

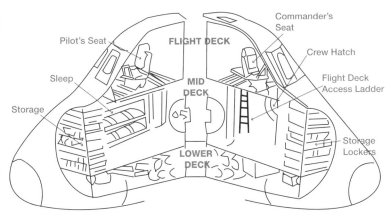

Cut-a-way view of orbiter's two-story crew cabin showing flight deck on top level and living quarters in mid-deck level

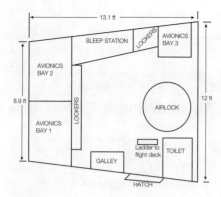

larger when projected on the wall or screen. This activity can introduce the relationship between the size of diagrams and the size of the objects or areas they represent.

DEVELOPING THE ACTIVITY

Have four volunteers stand at the vertices of a trapezoid to define the mid-deck living area, as shown on the diagram. First, have an estimation team decide how far apart and where the four students should stand for the living area to have the appropriate dimensions. Next, a measuring team actually measures the distances and, if necessary, adjusts the location of the students for the vertices.

Ask all students to sit along the sides of the trapezoid to form the boundary of the orbiter's living area. Finally, use masking tape on the floor to outline the area.

Compare the size of the orbiter's living area to such familiar areas as the size of a school bus, rooms in their homes, and the principal's office. One common benchmark for area is the size of a single bed. You may want to use a fitted single-bed sheet to help students compare the Shuttle's living area with the area in a child's bedroom.

Class Conversation

- How does the size of this living area compare with the size of our classroom?
- How many living areas do you think can fit in the classroom?
- How could we check our estimates?
- How many student desks can fit into the living area?

Remind students that crews of four to seven astronauts spend about two weeks together in the orbiter's living area. However, since they are in microgravity, they can move over each other. This explanation may help students understand the difference between square units of measure for flat surfaces and cubic units of measure for three dimensions. Even though the living area has about 160 square feet, the astronauts have about 2500 cubic feet of space when the flight deck and the mid-deck living areas are combined.

Part 2: Packing Up and Supplying the Ship

A mission into space takes careful planning. Once launched, the crew members cannot go home to get something they forgot. Ask students to work in small groups for about ten minutes to list what a crew member would want to take for a two-week trip. In a whole-class discussion, record the items suggested by the groups. Be sure to remind students of the limited amount of space.

Part 3: A "Day In Space"

NASA astronauts Coleman, Leslie, and Lopez-Alegria in a "sleep restraint" bunk bed

Ask the class to brainstorm what an astronaut's day in space might be like. Encourage them to make a daily schedule for astronauts, including 8 hours for sleeping, 4 hours for meals and hygiene, 9 hours for work, 2 hours for exercise, and 1 hour for relaxing each day. Their work includes conducting experiments; putting on space suits to do extravehicular activities (EVA), such as repairing satellites; cleaning the living area; and maintaining the orbiter. Students can make timelines or circle graphs to represent a day's activities or all the activities of a mission.

Then form crews of five to seven students to stay for a set period of time within the taped-off area that represents their orbiter living area. One whole school day is a suggested time. Make a class schedule so all students know when each team will have its "day in space."

Using its supply list, each crew must put all supplies and equipment needed for the duration of its mission into the living area before blast-off. The students need to cooperate with one another to get everything they need into the limited living area. Each student can pack personal items in a backpack or a grocery bag. Students may leave their orbiter only to take restroom breaks and to get their food from the school cafeteria. If conditions are appropriate, encourage students to return to their orbiter to eat their lunch.

Each crew makes a list of what crew members can do while spending their day in space inside the orbiter's living area. Suggested activities follow.

Astronauts can choose from more than seventy food items and twenty beverages. Some Space Shuttle missions have a galley with special serving trays, a convection oven, and a small dining area with a table and foot loops so astronauts can stay in one place while eating. Other missions do not have a full galley. On those missions, astronauts use a small food warmer.

- Crew members follow along with daily lessons, but to ask or answer questions, they must use a communications signal or device. Raised hands cannot be seen from Earth!

- Astronauts exercise within the living area when the class is at recess. Some activities, like jumping rope, do not work in microgravity. Doing bent-knee sit-ups with students taking turns holding one anothers' ankles, squeezing hand grips or exercise balls, and lying on their back with their feet in the air doing bicycle-peddling motions approximate microgravity exercises.

- Earth observations are an important part of Shuttle missions. Astronauts can "take pictures" of Earth from different points of view by drawing different physical maps of the world, for example, by placing the North Pole in the middle of the page rather than at the top.

- Experiments that measure living things in space can easily be role-played. For example, crews can measure classroom plants and animals—although scales do not work in microgravity! They can measure the number of heartbeats a minute both before and after exercise and record the data in their log books.

- Each NASA mission has its own patch, which is created by the crew. It contains the last names of the crew members and a visual representation of the mission. Each crew in your class can create its own patch to represent its "day in space."

- Living in microgravity has an effect on the cardiovascular system. The blood from the lower part of the body collects in the upper body. Students can simulate this experience with the following experiment.

  1. Wrap a ribbon or string snugly around the largest part of the calf of one member of the crew.

  2. With a washable marker, trace the top of the ribbon on the student's skin.

  3. Mark the spot where the end touches the ribbon, and measure the length in millimeters.

  4. Have the student lie down on his or her back with feet up against the wall or on a chair.

  5. At the end of 5 minutes, while the leg is still elevated, measure the calf at the same place and record the results.

Living Areas

NCTM Assessment Standards:
Assessment practices should shift toward integrating assessment with instructional activities.

Crew members can keep a flight log of their mission and can summarize the findings of their experiments. Encourage them to include diagrams showing the measurements used by the class to outline their orbiter living area in the classroom and their reflections about being confined with a team in such a small area.

Each crew should start its trip with a launch. Students not on the day's mission should role-play as members of ground control. As the day progresses, the astronauts can "transmit" the results of their experiments to ground control. Only verbal communication is allowed.

CLOSING THE ACTIVITY

At the end of the set time, ground control directs the crew to return to Earth. After the crew "lands," crew members should be debriefed about the mission. Ask students about any adjustments they learned to make. Encourage input from ground-control observers who have impressions about the astronauts' experiences.

As scientists, the crew has a responsibility to organize, display, and interpret the data collected during its experiments. Allow each crew time to share its findings with the class.

FURTHER EXPLORATIONS

- On cardboard from boxes, draw or paint different parts of the living area, such as the galley, lockers, hatches, airlocks, or sleeping pallets. Make a three-dimensional frame for the living area of the orbiter with rolled newspaper. Roll each sheet on a diagonal and secure with masking tape. Tape the rolls together to form the frame.

- Trace the entire area that the Space Shuttle covers. This activity can be done outside with chalk on a paved area or on a field with the liner that is used to mark baselines on a ballfield. Each area of the Shuttle can be marked and then compared with the size and location of the living area. The solid rocket boosters and the external fuel tank as they are positioned before launch can also be included.

- Modify the living area to be proportional to the students' size. The teacher should sit in a student's chair at the front of the classroom. By comparing how a teacher looks in a student's chair with how the teacher looks in his or her own chair, students can see that objects and areas can be sized differently. If the students are about 4 feet tall, use 4 feet for every 6 feet given on the diagram of the living area. With masking tape, make a new outline of the scaled-down size for comparison.

STS-73 pilot Rominger

Mission Mathematics: Grades K–6

MID-DECK FLOOR PLAN

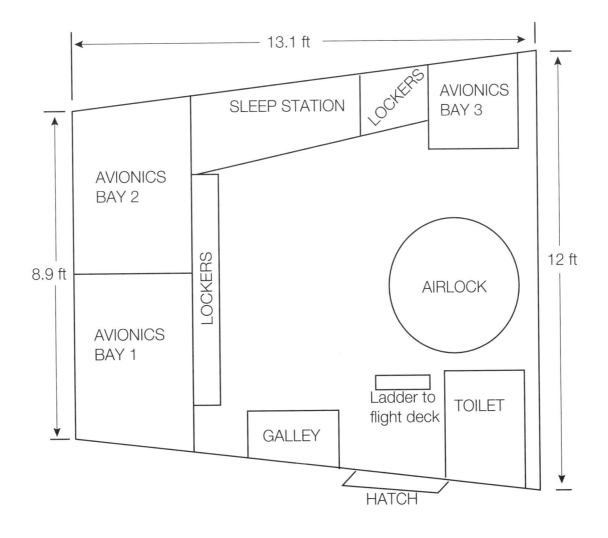

Living Areas

SPHERES IN SPACE

K 1 2 3 4 5 6

Students explore spheres by using balls and oranges. They make a pattern, or net, for a sphere with an orange peel. Students construct a sphere that they can personally occupy and compare it with the personal rescue enclosures (spheres) used by astronauts in their training. Students role-play rescue operations on a Shuttle mission.

NCTM Mathematics Standards
- *Measurement*
- *Estimation*
- *Geometry and Spatial Sense*
- *Patterns and Relationships*

Materials: *Several balls of different sizes; prescored, thick-skinned oranges; string; scissors; tape; tape measures, yardsticks, or metersticks; large pieces of paper, such as from a roll of butcher paper, or fabric (see part 3); three standard-sized play hoops (about 34 inches in diameter); duct tape; two bolts with wing nuts (optional); a globe (optional); and a world map (optional)*

Management Tip: *These activities can be used individually or as a unit over several instructional periods. Select activities appropriate for your students.*

INTRODUCTION

Space suits for Shuttle astronauts come in only three sizes: small, medium, and large. However, the astronauts' gloves are custom fitted. The upper and lower pieces of the space suit snap together with seal rings. The life-support system is built into the upper part. Usually only two space suits are supplied on a mission.

Personal rescue enclosures are in an experimental stage of development. These would be used to transfer crew members from one vehicle to another should the need arise. These enclosures are pressurized spheres, approximately 34 inches in diameter, that contain a life-support system and communications gear.

If it became necessary for crew members to move to another vehicle, two astronauts would wear the space suits and the other crew members would get into a personal rescue enclosure. Then the two astronauts in the space suits would move the occupied enclosures to the other vehicle by using a pulley-and-clothesline device or by towing.

PURPOSES

- To identify circles and spheres
- To make a net that forms a sphere
- To estimate and measure the diameters and circumferences of circles and spheres

Part 1: Which Are Circles? Which Are Spheres?
GETTING STARTED

Ask the class to sit in a circle. Have them reflect on the properties of a circle by asking how they know that they really are sitting in a circle. Ask a few students to estimate the location of the center of the circle. Then have students use a string to show that everyone is about the same distance from the center.

As you show students different-sized balls, ask them to explain the difference between the ball and a circle. If they cut a ball of clay through the center, what shape would they see?

Introduce or review the meaning of the word *sphere*. To help students differentiate between a sphere and a circle, have them work together to make a two-column list of objects that are three-dimensional spheres or that are circular, but flat. Encourage students to continue adding to the list each day as they encounter or remember other objects.

Ask students to think about how to make a paper pattern that could be cut out and taped together to make a sphere.

DEVELOPING THE ACTIVITY

As you hold up an orange, tell the class that you have a special way to peel the orange. Slice a half-dollar-sized circle off the top, or stem end, of the orange. Hold it up to the class and ask students to identify the shape of the cross section. Cut through the skin of the orange from the top to within 1/2 inch of the bottom

Space-suited Robert Williams does an evaluation run with a personal rescue enclosure.

at eight to ten evenly spaced intervals around the orange. Leave an uncut, half-dollar-sized piece at the bottom. One by one, slowly peel each segment of skin down to the bottom of the cut. The orange peel should remain connected at the bottom. It will look something like a flower, with the fruit of the orange sitting in the center of the "petals."

Distribute oranges with the skins precut, and have students carefully peel their oranges in the same way. Students can remove the fruit, but make sure that they save the peel intact. Students can share the fruit during a classroom break.

CLOSING THE ACTIVITY

Ask students to reshape the sphere by holding the peel together. Students may want to help one another. Then students should trace on paper the flattened peel to make a pattern, or net, for a sphere. They cut out their pattern and tape the pattern together to form a rough model of a sphere. Encourage a discussion of their experiences by asking students what surprised them about the shape of the pattern for the sphere. Have them describe what the pattern for the sphere looks like when it is flat.

Part 2: How Big Is Your Rescue Sphere?
GETTING STARTED

Encourage students to share their knowledge of, and experiences with, space suits. As you show an illustration of a space suit, ask why astronauts need space suits. What special equipment must a space suit have for an astronaut? [breathing apparatus, micrometer protection, proper pressure, temperature-control devices, communication devices, drinking water, food, waste-collection device, and electrical system]

Explain that all members of a crew do not go outside the Shuttle on routine space flights. Since space suits are very expensive, NASA no longer equips each Shuttle astronaut with a personal space suit. Normally only two suits are available on a Shuttle mission.

To prepare for an emergency for which the crew has to be moved outside the spacecraft, NASA is developing a special sphere called a *personal rescue enclosure*. Show a picture of NASA's personal rescue enclosure. Explain that this sphere contains a life-support system and communications gear. It is deflated to save space when it is stored on the orbiter.

Begin a discussion by asking students how large they think the sphere needs to be when it is inflated for an adult astronaut to fit inside.

Students may respond by showing the size with their arms, by comparing it with familiar objects, or by estimating the measures of the circumference and the diameter.

NCTM Teaching Standards: *It is important that students are engaged in worthwhile mathematical tasks. One important aspect of this activity is that students are challenged to begin developing a coherent framework to contrast two- and three-dimensional figures.*

DEVELOPING THE ACTIVITY

Divide the class into small groups. Students experiment to find how large a sphere they would need for a personal rescue enclosure. Each student takes a turn rolling his or her body into as small a spherelike shape as possible. The other students wrap a string around the student to determine, measure, and record the circumference of a sphere that fits this astronaut. Then the student ties or tapes the ends of the string together to form a circle and traces the circle on large paper. Students cut out their circles, measure and record the diameters, and personalize the circles with self-portraits.

CLOSING THE ACTIVITY

Make a circle with a 34-inch diameter. Let students compare the sizes of their circles with the size of your circle, which has about the same circumference as NASA's personal rescue enclosure.

Teaching Note: *If some students in your class are large or small for their age, you may want to change the measuring phase of the activity to avoid embarrassing them. You could ask adult volunteers to be models.*

Spheres in Space

Students can order their circles by the size of the diameters. Finally, ask students to predict whether the order will change if they use the circumferences instead. Have them verify their predictions and summarize their findings.

You may want to suggest that students design a way to display all their circles in the classroom.

Part 3: Into Our Rescue Sphere
BEFORE THE ACTIVITY

Use three large, plastic play hoops with diameters of approximately 34 inches to make a sphere that is about the same size as NASA's personal rescue enclosure. To make the frame, join the three play hoops by using a 1/4-inch drill bit to drill carefully through each hoop at opposite poles. Join the three hoops at these points with bolts and wing nuts. Tape cotton balls around the wing nuts to protect students. Spread open the hoops until they are at approximately 60-degree angles to each other. An alternative method is to use duct tape to attach the play hoops at opposite poles.

Once the frame is constructed, cover five sections with the large paper or fabric. Use wide masking tape or duct tape to keep the covering in place.

GETTING STARTED

Review illustrations of a space suit and NASA's personal rescue enclosure and the circles from their own personal "spheres." Ask students to describe how they might feel if they were in a sphere like NASA's personal rescue enclosure for a long period of time. What should be inside the enclosure to help an astronaut being rescued from one spaceship to another?

DEVELOPING THE ACTIVITY

Explain that you have made a sphere that is the same size as NASA's personal rescue enclosure.

Class Conversation

- What shapes do you think were joined to make the covering for this sphere-like shape?
- About how big is our personal rescue enclosure?
- Why do you think NASA engineers decided to use a sphere instead of another shape, such as a cylinder?

Ask the students to measure and record the circumference and diameter of the class rescue sphere. Students should compare the diameter of the sphere with familiar lengths, such as their arm span, their height, the height of their desk, the diameter of a doorknob, the teacher's height, and a school bus seat.

CLOSING THE ACTIVITY

Encourage students to take turns sitting inside the play-hoop sphere to get the feeling of being in a rescue sphere. Each child can choose whether or not to close the sixth flap. Students can take turns sharing their feelings about being in the sphere.

You may want to keep the personal rescue enclosure in the classroom for a few days for students to revisit.

FURTHER EXPLORATIONS

- Some students may be ready to compare the size of their sphere with that of an adult astronaut, using ratios.
- Compare volumes of different spheres. Use a hemispherical-shaped container,

---

Note to Intermediate-Grades Teachers: *A small group of students may be able to construct the sphere with little adult supervision. If bolts and wing nuts are used to secure the play hoops, an adult should make the holes with the drill.*

Assessment Notes: *Ask students to draw what the "skin" of a basketball would look like if it were "peeled" and flattened.*

Students should be able to estimate whether the circumference and diameter of their circles are greater than, the same size as, or less than those of the circle that has about the same-sized circumference as NASA's personal rescue enclosure.

Mission Mathematics: Grades K–6

such as a bowl. Fill the bowl twice—once with golf balls and once with tennis balls. Compare the size of the balls. Compare the number of balls of each type that fit in the sphere. Students should conclude that more of the smaller balls fit in the hemisphere.

- Students can design a personal rescue device with a different shape. Ask them to estimate the dimensions and share their designs.

- After sitting inside the paper sphere, students can write a story about how it feels to be in the sphere while being rescued.

- Show the students a map of the world. Compare it with a world globe. How do cartographers show the surface of a sphere on a flat piece of paper?

School-Home Connection: *Students can write stories or draw pictures to share with their families their experiences inside the sphere. They can include a description of the sphere's size.*

Spheres in Space

DESTINATION: SPACE STATION

K 1 2 3 4 5 6

NCTM Mathematics Standards
- *Geometry and Spatial Sense*
- *Patterns and Relationships*
- *Number Sense and Numeration*

Materials: *Large, plastic play hoop, approximately 34 inches in diameter; masking tape; chalk or string; graph paper; measuring tape or yardstick*

Astronauts working on a satellite

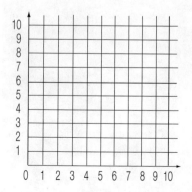

Students maneuver their Space Shuttle or manned maneuvering unit on a coordinate grid by following directions given as ordered pairs of numbers. They simulate orbiter and space-station conditions of working together in a very small space to complete their mission successfully.

PURPOSES

- To follow a path on a large coordinate grid using ordered pairs
- To identify patterns, relationships, and geometric shapes on a coordinate grid
- To coordinate the movements of two or three crew members
- To create problems involving points on a coordinate grid

INTRODUCTION

Manned space flight has always involved people working together in extremely small spaces. Even though the Shuttle orbiter looks very large to us, the space in which the astronauts live and work is very limited. This situation is also true on the International Space Station, where the living and working modules are about the same size as a school bus, although the entire space station will be the size of about 1.5 football fields.

GETTING STARTED

Show the class the graph paper and ask how they can make on the floor a grid that looks like the paper. Help a measuring team use chalk, string, or tape to mark off a large grid either outside on a court or field or inside on a floor. Square tiles on a classroom floor present a great pattern for a grid, although they are too small to use as unit squares for this activity. Talk about the need to mark parallel lines and equal intervals on both the horizontal axis, or *x*-axis, and on the vertical axis, or *y*-axis. The parallel lines of the grids should intersect to form congruent square regions.

After all students have had an opportunity to learn about the *x*-axis, the *y*-axis, and the origin—the intersection of the two axes—of the coordinate grid, introduce or review how to name points on the grid with ordered pairs. The first number in each pair represents the number of spaces moved horizontally from the origin and the second number always represents the number of spaces moved vertically. For example, to locate the point named by the ordered pair (3, 8), start at the origin and move to the right three spaces and up eight spaces. Students familiar with, or ready to learn about, integers can locate the ordered pair (–5, –2) by starting at the origin and moving to the left five spaces and then down two spaces.

In each of the following missions, the teacher should prepare the initial coordinates and problems to be solved.

Mission 1: Satellite Rescue

This activity begins with students using a play hoop (*a*) to gain concrete experiences with the grid system and (*b*) to work together in small spaces. Invite three students at a time to step into a large plastic play hoop—their spaceship.

The size of the students should be considered when deciding whether to use two or three students. Students are to move on the coordinate grid without holding the play hoop with their hands. In each team of astronauts, one student is designated the pilot of the mission. Only the pilot's feet determine the location of the spaceship.

Astronauts need to leave their orbiter to rescue a satellite and bring it safely inside the cargo bay. The team receives directions for capturing a satellite from ground control. The directions are in the form of ordered pairs. Students record their team's path by attaching a string to their play hoop and letting it flow as they move to the designated points. Secure the string to the grid with masking tape at each point. The following sample mission uses integers and all four quadrants:

- Your orbiter is located at (–5, 2). To reach a satellite that needs repair and return it to the cargo bay, program your MMU (manned maneuvering unit) to follow this route: Go to position (2, 3), and pick up the satellite at (2, –2). Next, go to (–5, –2); finally, go to (–5, 2).

- Now that you have the satellite in the cargo bay, look at the path you took. What geometric shape does your path follow?

Mission 2: Through the Asteroids

Teams of astronauts plan and test a safe path from the origin, (0, 0), through the asteroids, to the space station located, for example, at (10, 8). To help students steer clear of the asteroids, the locations of the dangerous objects are given to the astronauts as ordered pairs, (x, y). For example, you could place asteroids at (3, 2), (7, 2), (8, 4), (3, 6), (6, 8), (10, 6), (6, 8), and (10, 9).

Two astronauts on each team make up and file their flight plan of ordered pairs for a safe path through the asteroids. Then they test their path by directing a third team member to follow their flight plan around the grid. Encourage teams to use the fewest number of points to reach the space station without hitting an asteroid.

Mission 3: Docking with Mir

Astronauts must dock their orbiter with the Russian space station, *Mir*. The coordinates of the locations of the orbiter and *Mir* are given as ordered pairs, (x, y). Each firing of the thrusters propels the orbiter two horizontal or vertical spaces. Students must decide how many times to fire the thrusters to reach *Mir*.

Suppose, for example, that the Shuttle is located at (4, 2) and *Mir* is located at (0, 0). How many times must the thrusters be fired? In this example, students are working with even numbers; thus, the coordinates chosen need to be even numbers. To increase the difficulty level of this activity, the pilot can choose to use long burns to move two spaces or short burns to move one space. Which pilot can reach *Mir* by firing the fewest number of burns and, therefore, use the least amount of fuel?

FURTHER EXPLORATIONS

- As students gain proficiency, they should be encouraged to create problems for their classmates to solve.

- Students can place four checkerboards together. Problems can be solved using counters or checkers to locate the orbiter, asteroids, satellites, and the space station on the grid.

NCTM Teaching Standards:
Teachers must decide when and how to use mathematical notation and language. Primary-grades teachers will want to limit all points to quadrant I, using whole numbers only. Students who are ready to learn about negative numbers can move into quadrants II, III, and IV and use coordinates with integers.

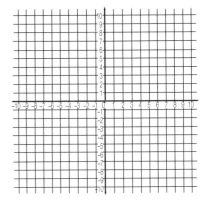

NCTM Assessment Standards:
Students should be made aware of expected outcomes of their learning experiences. Your class can celebrate one another's successes when they locate a point on a coordinate grid when given the coordinates in (x, y) form and name the geometric shape formed by a set of vertices given in (x, y) form.

Destination: Space Station

SPACE SCIENCE

For centuries, humans from many cultures have looked to the heavens to observe patterns that might give clues about who we are, how we got here, and where we are going. Copernicus, Galileo, and Kepler are a few of the observers of the heavens who gave humankind new insights into the design and composition of the universe. NASA seeks to answer fundamental questions about the universe just as earlier astronomers did. However, NASA scientists have space-based telescopes to observe the heavens; space probes, orbiters, and landers to explore the planets; and Earth-orbiting satellites and deep-space missions to study the solar system.

As we look to the future, NASA's space-science program helps keep the United States at the forefront of science. Exploring beyond Earth's orbit, NASA seeks answers to fundamental questions about the origin of the universe, the way the universe works, the relationship between the Sun and Earth, and life in the universe. Well-planned scientific missions related to space science continue to enlighten us about these questions.

✦ Hubble Space Telescope and Compton Gamma Ray Observatory data have enabled scientists to make discoveries about the size, age, and evolution of the universe, including black holes, quasars, and its "missing" dark matter.

✦ Observing the impact of the comet Shoemaker-Levy 9 with Jupiter through land-based and space-based telescopes allowed scientists to gain insights into the nature of craters on the planets, on the Moon, and even on Earth.

✦ Data from the Cosmic Background Explorer have confirmed the big bang theory about the origin of the universe.

✦ Unexpectedly, the joint European Space Agency/NASA *Ulysses* solar spacecraft found that the Sun lacks a magnetic pole.

✦ The *Galileo* spacecraft showed on a first-ever flyby of asteroids that a moon orbits the asteroid Ida.

✦ During the *Astro-1* mission, the Ultraviolet Imaging Telescope obtained a large number of images, including clusters of young, hot, massive stars; globular clusters containing old stars, some of which are unusually hot; spiral galaxies rich with star-forming activity; and smaller irregular galaxies that can experience sudden bursts of star formation.

✦ *Voyager 2* flew within 5000 kilometers of Neptune while it was the most distant planet from the Sun. The images from *Voyager 2* solved many of the questions about Neptune. The rings of Neptune are complete rings. Neptune has several large dark spots, reminiscent of Jupiter's hurricane-like storms. The magnetic field of Neptune, like that of Uranus, is highly tilted (47 degrees) from the axis of rotation. *Voyager 2* found six of the eight satellites of Neptune.

The information gained from the study of space science is varied and rich. The missions underway are providing data sets for many scientists to evaluate. We can expect that information gained from these data will encourage even more discoveries about the nature and origin of the universe. The true challenge will be to see the underlying patterns of data from so many sources.

AN INTRODUCTION TO SPACE SCIENCE ACTIVITIES

When you look into the night sky, you are looking into a history of the universe. The sunlight that shines on us is 8.5 minutes old when it reaches Earth. Most of the stars we see in the sky are hundreds or thousands of light-years away. The Andromeda galaxy, a mere 2.2 million light-years away, is truly a next-door neighbor. The activities of this section are designed to help your students learn about the missions of NASA that help us learn more about our solar system and our universe.

Future NASA probes that will eventually reach Saturn and Pluto include *Cassini,* a joint United States–European mission to Saturn and its moon, Titan. It is to be launched in October 1997 and should arrive on Saturn in June 2004, after completing flybys of Venus and Jupiter. A Pluto fast flyby is being planned for a small spacecraft. If launch is achieved in 1999 or 2000, this probe could encounter Pluto and its moon, Charon, around 2006 to 2008.

All activities of the space-science section integrate the four NCTM process standards: problem solving, mathematical reasoning, mathematical communication, and mathematical connections. Each activity also incorporates several of the NCTM mathematics standards.

The Hubble Space Telescope is only one of the tools used by scientists to explore our universe. This telescope was launched in April 1990 by the crew of the Space Shuttle Discovery. *The first servicing mission was completed in December 1993 by the crew of the Space Shuttle* Endeavour.

The second servicing mission was completed in February 1997 by the crew of STS-82 on the Space Shuttle Discovery.

The following brief outline describes the "Space Science" activities and the associated mathematics.

Activity Name	Mathematics	Grade Levels
Probing the Planets	Fractions, mixed numbers, decimals, measuring	K–6
How Much Does the Milky Weigh?	Ordinal numbers, sequencing, patterns	K–2
Gravity and Weight	Congruent figures, translations, reflections, and rotations	3–6
Mission to Mars	Probability, data analysis, patterns	K–6
Journey to Jupiter	Probability, data analysis, patterns	K–6

Other sources of information of space science and the related technologies are available through NASA's home page on the Internet. For more information about NASA's materials for teachers, see appendix A.

PROBING THE PLANETS   K 1 2 3 4 5 6

NCTM Mathematics Standards
- *Patterns and Relationships*
- *Number Sense and Numeration*
- *Statistics*
- *Estimation*
- *Measurement*
- *Fractions and Decimals*

In these problem-solving activities, students create models of our solar system. Primary-grades students learn that some planets are larger than others, and they make a model to show the order of the planets from the Sun. Intermediate-grades students model the relative sizes of the planets and their distances from the Sun.

PURPOSES

For all students:

- To model representations of the planets' sizes
- To compare and order whole numbers
- To interpret data to discover patterns and relationships
- To model representations of the planets' order from the Sun

For intermediate-grades students:

- To experience meters and kilometers as units of measure for distance
- To estimate planet sizes and distances from the Sun
- To read, interpret, and sequence data about the planets by using tables
- To choose appropriate graphs for displaying data
- To compare and order decimals
- To model representations of the planets' distances from the Sun
- To use diameters to measure spheres and circles

Primary-Grades Activity: Solar System Kabobs

For students who are not ready for standard units of measure or place-value concepts, a simple ordering activity is appropriate. Students make planet kabobs with food items on bamboo skewers to model the order of the planets from the Sun.

GETTING STARTED

Use the given planet sizes in the table on page 75 to sort the planets into three basic sizes—large, medium, and small. The following readily available foods are suggested to represent the relative sizes of the planets.

Saturn and its beautiful rings are of interest to NASA scientists. The spacecraft Pioneer traveled over 2 billion miles to take close-up pictures of Saturn.

Materials: *Bamboo skewers; food items to represent the planets and the Sun; paper plates or paper napkins; a set of index cards with planets' names and ordinal positions from the Sun*

Mercury: 1/2 gum drop	Saturn: dried prune
Venus: jelly bean	Uranus: large red grape
Earth: green grape	Neptune: minimarshmallow
Mars: 2/3 gum drop	Pluto: raisin
Jupiter: dried apricot	

DEVELOPING THE ACTIVITY

Help students make cheese-toast Suns, which they place on a plate or napkin. Students thread their "planets" onto bamboo skewers in order from the Sun, beginning with Mercury. As each planet is added, talk about the planet's size and ordinal position from the Sun.

CLOSING THE ACTIVITY

When the skewers are complete, students place their planet kabob next to the Sun on the plate. As students eat their planets and Sun in reverse order, they can practice counting backward from 9. You may want to ask about other relationships.

Class Conversation

- Which planet would be hottest? Coldest? Why?
- Which planet is largest? Smallest?
- What are some other large planets? Other small planets?
- What are some planets that are medium-sized?

FURTHER EXPLORATIONS

Write the names of the planets on index cards. After students have modeled the planets with the kabobs, they can order the cards with respect to the ordinal positions of the planets from the Sun.

Intermediate-Grades Activity:
Welcome to Our Solar System

BEFORE THE ACTIVITY

Prepare the sets of index cards for each group. If your class is not familiar with meters, kilometers, and diameters of spheres and circles, introduce these concepts before beginning this activity. Suggestions follow in Getting Started.

The following foods are appropriate for representing the planets.

	Diameter (in km*)	Suggested Model
Sun	1 400 000	large yellow ball or balloon
Jupiter	143 200	pumpkin
Saturn	120 000	acorn squash
Uranus	51 800	large onion
Neptune	49 500	large tomato
Earth	12 800	cherry tomato
Venus	12 100	walnut
Mars	6 800	large cranberry
Mercury	4 900	peanut
Pluto	2 300	small pea

* rounded to the nearest 100 km

Materials: *For each group, two sets of index cards, one with the planets' names and sizes and another with their names and distances from the Sun; string or twine at least 59 paces long*

Optional Materials: *Suggested books:* Is a Blue Whale the Biggest Thing There Is?, How Much Is a Million?, The Magic Schoolbus Lost in the Solar System; *paint, roll paper, compass and ruler; clothesline and clothespins; appendix C*

NCTM Assessment Standards: *Teachers must continually assess and compare students' understandings with the goals they are expected to achieve. An informal class discussion can be used to assess students' prior knowledge of our solar system.*

GETTING STARTED

To help students begin thinking about relative sizes and distances, read them a book such as *How Much Is a Million?*

Ask students to estimate the size of Earth and the distance from Earth to the Sun. Review, or introduce, the meter and kilometer as units of measure. Discuss the relation between meters and kilometers (1000 meters = 1 kilometer).

Review, or introduce, how circles and spheres are measured by their diameters. If appropriate for your class, use skewers or toothpicks and fruit from the model of the solar system to show how to measure the diameters of a sphere.

Discuss distances between locations. If appropriate, take a walk about 1 kilometer in length. Have students predict where 1 kilometer, or 1000 meters, will be on the path of their walk. Use such benchmarks as 10 meters, 20 meters, 50 meters, or 100 meters to measure familiar paths in the school, for example, the

Probing the Planets

Cross-Curricular Connection: *In physical-education classes, students can participate in field and track events that involve distances measured in meters and kilometers.*

Graphing Tip: *When the range of a data set is extreme, it may be helpful to group the data into equal-sized intervals and make a histogram rather than a bar graph.*

distance from the classroom to the cafeteria. Then discuss greater distances between locations—in the neighborhood of the school; in the local city, town, or county; in the state; and in the United States and on Earth.

Give students the opportunity to revise their estimates of the size of Earth and the distance from Earth to the Sun. Record their estimates in a class chart. Then, discuss appropriate types of graphs they can make to display their data. Ask students to make two different graphs—one to show their estimates of the size of Earth and the other to show their estimates of the distance from Earth to the Sun.

DEVELOPING THE ACTIVITY

At this time, give students opportunities to learn more about the different planets from a variety of available materials, such as computer software, electronic networks, books, and videos, such as *The Magic Schoolbus Gets Lost in Space*. See your school's or district's media specialists for resources.

With the class, make a data chart of the planets in alphabetical order. Include columns in the chart for recording the size (diameter in kilometers) and distance from the Sun (in millions of kilometers). See appendix B for data.

Students can practice reading and comparing large numbers.

Hold up the object you have selected to represent our Sun, such as a very large yellow ball or balloon. If the class has read *Is a Blue Whale the Biggest Thing There Is?*, you may want to revisit the benchmark relating the size of Earth to that of the Sun.

Mix up the nine objects selected to represent the planets, and share them with the students. Do not tell the students which planet each object represents. Display the objects so all students can observe and touch them and make comparisons. Tell the students they will use the objects to model our solar system. Then, present this problem.

> The first step in making a model of our solar system is to use our chart with the sizes of the planets to decide how to match each object to a planet. Work with your group to decide which object should be used to represent each planet.

Problem Solving and Cooperative Learning: *Working together in small groups can provide nonthreatening, enjoyable experiences that build your students' confidence in their problem-solving abilities.*

Encourage all members of the groups to discuss their understanding of the problem, to plan their strategy, to work together using the agreed-on strategy to solve their problem, and to reflect on their solution to see if it is reasonable.

When all groups have a solution, have each group present its order for the objects by size, and place the appropriate index card with the planet's name and size next to each object. Ask each group to share its strategy for solving the problem, its results, and the reasoning behind its decisions.

Assist in recording the results for each group in a class chart. Allow time for the class to reflect on and discuss the different strategies and results.

For the next section of the activity, modeling the planets' distances from the Sun, you may want to take students, the objects representing the planets and the Sun, and the two sets of index cards outside or to an indoor activity area, such as a hallway or a gymnasium.

Writing Connection: *Remind students to record notes about their problem-solving process. They may want to use these steps:*
- *Understand*
- *Plan*
- *Solve*
- *Look back*

The following data are ordered according to distance from the Sun in millions of kilometers. Do not share this chart with students at this time. Have students match the objects with the index cards indicating the planet's name and size. Then, present this problem.

For the next step in modeling the solar system, we need to place the planets in the correct order, showing their distance from the Sun. Work with the members of your group and use our class data chart.

	Distance from Sun (in millions of km*)	Paces
Mercury	58	about 0.5
Venus	108	1
Earth	150	about 1.5
Mars	228	about 2.5
Jupiter	778	almost 8
Saturn	1427	about 14.5
Uranus	2871	almost 29
Neptune	4497	almost 45
Pluto	5913	59

*rounded to nearest 1 million km

Language Connection: *When the class has ordered the planets according to distance from the Sun, encourage them to use the first letter of each planet's name to create their own mnemonic to help them remember the order.*

When first discussing the distance data, focus on the largest common place value, using the word *millions* instead of recording and reading all the zeros. For example, say "Mercury is about *58 million kilometers* from the Sun." Then have students order the distances.

Next, have students compare the distances. Ask them to identify any relationships they see. For example, students might note that Mars is about 228 million kilometers from the Sun, which is more than twice the distance for Venus, 108 kilometers.

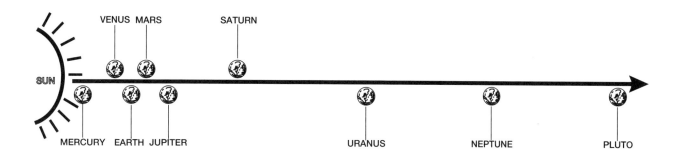

So students can show relative distances in their completed model, begin by helping them analyze the distance data in terms of fractions and estimation. Have students use their data about the position of the planets to write questions that involve the use of the fractions 1/2, 1/3, or 1/4. Some sample questions follow.

- Which planet has a distance from the Sun that is about 1/2 the distance from the Sun to Pluto? About 1/4 the distance?
- Which planet is farther from the Sun than about 1/2 the distance from the Sun to Pluto?
- Which planet is located about 1/3 the distance between the Sun and Jupiter?

Give students twine or string and ask them to pace off the distance between the Sun and Pluto with 59 paces, since Pluto is 5900 million kilometers from the Sun. Have them place the object modeling the Sun at the beginning point on

Problem-Solving Strategy: *Supplying students with the number of paces for the greatest distance and asking them to find the number of paces for the other distances give them the opportunity to use the work-backward strategy to solve a problem.*

Probing the Planets

the string and the object representing Pluto at their ending point. Then pose the following problem.

> The distance from the Sun to Pluto is 5913 million kilometers, or about 59 hundred million kilometers. If we use 59 paces to show this distance to Pluto, how many paces should we use to show the distances of the other planets from the Sun?

CLOSING THE ACTIVITY

After all groups have shared their solutions, encourage students to look for patterns and relationships in their data and model. Help students record their conjectures. Share students' questions that help them think about other relationships in the data.

Students should reflect on the activity and recall the three methods they used to order the planets: by alphabetical order, by size, and by distance from the Sun. Ask students whether everyone had the same solutions for the three problems. Finally, ask students what other things about the planets they would like to learn.

FURTHER EXPLORATIONS

- Students can place clothespins labeled with the planets' names along a clothesline to show the planets' relative distances from the Sun.

- Give each group of students about 2 or 3 meters of 36-inch-wide dark roll paper on which to paste circles representing the Sun and planets of their solar-system model. Have them draw circles that are proportional by comparing the planet diameter data, measuring diameters with rulers, and drawing the circles with a compass. Students may need some hints, such as "Draw a circle with a diameter of 5 millimeters to represent Pluto's 2300-kilometer diameter and a circle with a diameter of 12 millimeters to represent Mercury's 4900-kilometer diameter." Students can make the other circles by referring to the given diameters for Pluto and Mercury as the benchmarks. Have them use fractions and estimation to place the circles in order according to the distance from the Sun.

- Introduce another unit of measure, the *astronomical unit* (AU)—the distance between Earth and the Sun, about 150 million kilometers, or about 93 million miles. Challenge students to use calculators to find the distance of each planet from the Sun in astronomical units. For example,

$$\text{Pluto} \rightarrow \frac{5913}{150} \approx 39.4.$$

Pluto is about 39.4 AU from the Sun.

- Organize the class into research groups. Help students develop criteria for their research and presentations. Have them work together to create a chart of their combined research, using such categories as rotation period, period of revolution, mass, or atmosphere or temperature. You can incorporate writing assignments, such as creating a travel brochure for one of the planets.

Algebraic Thinking: *Proportional reasoning is an important building block for success in algebra.*

Assessment Note: *Ask students to record in their journals their thoughts about problems presented in this activity—the strategies and solutions used, the groups' different methods of working, and the difficulty of the problems.*

HOW MUCH DOES THE MILKY WEIGH?

K *1 2* 3 4 5 6

Students measure the weight of a miniature candy bar with nonstandard units of measure in a pan balance. Then they compare the relative weights of individual objects.

PURPOSES

- To measure the heaviness of a miniature candy bar with nonstandard units of measure
- To equalize weights using a pan balance scale
- To count and graph the number of objects used to measure weight
- To compare the relative weights of objects by interpreting a graph
- To explore informally the transitive property of equality: if $a = b$ and $b = c$, then $a = c$
- To explore, model, and communicate inequalities

INTRODUCTION

Our universe is named the Milky Way. Even though our solar system with its nine planets is very, very big, it is just a very, very small part of the Milky Way. NASA space scientists are learning about our solar system and our universe. One thing about which they are very curious is gravity. Gravity is one of the forces of the universe. Gravity is not the same intensity on all the planets of our solar system or in other parts of the Milky Way.

As NASA scientists travel into space, they perform experiments to learn more about how different gravitational forces affect people, plants, and materials in solid, liquid, and gaseous forms.

The experiments on an orbiting Space Shuttle are called microgravity experiments because the level of gravitational force is low. These experiments can last as long as the Shuttle mission lasts, usually about fourteen to seventeen days. Now that our astronauts are working and living in the *Mir* space station, these very important microgravity experiments can last for months.

Our understanding of the effects of gravity begins with measuring weights of objects. The act of balancing sets of objects on two sides of a pan balance enables students to experiment with the effects of gravity. Students do not need knowledge of units of measures, such as grams and pounds, to experiment with gravity with a pan balance.

GETTING STARTED

After sharing appropriate introductory materials with your class, ask students what words they know that sound alike but have different meanings and different spellings. Then discuss the language link between the name of the activity and the name of our universe.

Invite your students to pretend that they are NASA scientists who are measuring the weight of objects here on Earth. Ask them to think about how heavy they think a miniature candy bar is. If necessary, help them compare it with other familiar objects, such as paper clips, pencils, and crayons. Record their estimates in a class chart. Then ask how they can use a pan balance scale to measure the weight.

NCTM Mathematics Standards

- *Measurement*
- *Estimation*
- *Number Sense and Numeration*
- *Patterns and Relationships*

Materials: *Miniature candy bars; pan balance scale; a variety of small objects, such as paper clips, erasers, connecting cubes, paper fasteners, and plastic teddy bear counters; graph paper or large chart; crayons or markers; graphing software (optional)*

On 26 September 1996, Shannon Lucid returned to Earth in the Space Shuttle Atlantis *after spending a record 188 days in a microgravity environment.*

NCTM Teaching Standards: *Students should be able to use tools, including technology, to explore mathematics.*

If a computer and graphing software are available, students can easily explore different ways of displaying their data on the computer in picture graphs, pictographs, or bar graphs.

Algebraic Reasoning: *Although it may seem obvious or trivial to adults, the discovery of the transitive property of equivalence is a very big step for young children. They are exploring experiences that can help build a foundation for learning the algebraic property that if a = b and b = c, then a = c.*

Management Tip: *This activity is appropriate for a learning center for partners or small groups. After all students have had an opportunity to experiment and collect their data, you may want to conduct a whole-class discussion for displaying and interpreting the data.*

DEVELOPING THE ACTIVITY

Students should experiment with the scale by placing their candy bar in one pan of the scale and then placing enough of one type of object in the other pan to balance the weight of the candy bar.

As young students balance their candy bar with different sets of objects, an adult or older student may need to help them record their results.

To make a bar graph for the number of objects needed to balance the candy bar, students can place pictures of objects, or write names, across the bottom of a piece of graph paper or a chart. Then above each object they should color in the number of squares equal to the number of objects it took to make the pans balance. Proceed in this manner with all the sets of objects used in measuring the candy bar.

Students can compare the weights of the equalized sets of objects to see if they balance one another. To demonstrate, place the set of paper clips that balanced the candy bar on one pan of the balance and place a different set of objects that also balanced the candy bar on the other pan. Once everyone is convinced of the equivalence of the weights of the sets of objects, try to elicit explanations of how to draw a picture of the scale that shows the equivalent weights and why many paper clips but only a few plastic teddy bears are needed to balance the candy bar.

Students can role-play their discoveries of equivalent weights by extending their arms out from their sides and holding their hands palms up to simulate the pans of the balance. After several students have had opportunities to show their findings, ask students to think about what would happen to their scale if objects were added to, or taken away from, one hand.

Ask them whether the pans on their "scale" should balance or tip to the side. Help students decide which pan is heavier and which side of their "scale" should tip to one side. Ask students to think of ways to make the scale balance again. Continue this line of reasoning by asking how to make their "scale" tip in the other direction.

CLOSING THE ACTIVITY
Class Conversation

Ask students to use their graphs to reflect on questions like the following:

- Which type of object requires the most pieces to balance your candy bar? The least?

- How many paper clips does it take to balance with your candy bar? How many teddy bears? If you put that number of clips in one pan and that number of bears in the other pan, would the pans balance? Why?

- If you put a candy bar in one pan and three cubes in the other pan, will the pans balance? Which pan will be lower? Higher?

- If two candy bars are in one pan, how many cubes should you put in the other pan to make it balance? If there are three candy bars? If there are four candy bars? What pattern do you see?

ASSESSING THE ACTIVITY

Observe students to see which students can balance the sets of objects with the candy bar and then count the number of objects whose weight is equal to the weight of the candy bar.

Can students make a correct bar graph with their results? Can students interpret their results from looking at their graph?

FURTHER EXPLORATIONS

- When your students are ready, record the equivalencies and inequalities in number sentences.

- Extend students' reasoning to explore addition and multiplication relationships.

- Introduce standard units of measure and other types of scales.

- Encourage students to develop a whole new system of measurement, such as the Andromeda galaxy measurement system, which could be set up as a different base system.

RELATED ACTIVITIES IN MISSION MATHEMATICS

Students practice making picture graphs, pictographs, and bar graphs in "Covering an Airplane." They explore size and distance from the Sun in "Probing the Planets" and surface gravities relative to Earth's surface gravity in "Gravity and Weight."

How Much Does the Milky Weigh?

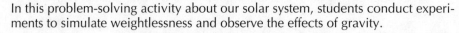

GRAVITY AND WEIGHT   K 1 2 3 4 5 6

In this problem-solving activity about our solar system, students conduct experiments to simulate weightlessness and observe the effects of gravity.

NCTM Mathematics Standards

- Patterns and Relationships
- Number Sense and Numeration
- Estimation
- Measurement
- Fractions and Decimals
- Statistics

Materials: *For each group: base-ten materials—flats, longs, and cubes; ankle weights, wrist weights, or both; two sets of index cards, one with the planets' names and sizes, another with the names and surface gravities relative to Earth's surface gravity; lettuce spinner and small objects, such as counters or connecting cubes; the book* Almost The Real Thing: Simulation in Your High-Tech World *(optional)*

PURPOSES

- To model decimal numbers
- To compare and order whole numbers and decimals
- To read, interpret, and sequence data about the planets using tables
- To interpret a glyph to discover patterns and relationships

GETTING STARTED

Informally discuss the concepts of weight and mass. Ask students what they know about astronauts' experiences of weightlessness as they orbit Earth on the Shuttle missions. Emphasize that even though the astronauts are weightless, they still have the same mass. The weight of an object is different from its mass. Weight is produced by gravity.

The gravitational force on Earth is about 6 times greater than on the Moon. An object that weighs 1 pound on the Moon weighs about 6 pounds on Earth. If the gravitational force on Jupiter is about 2.5 times greater than on Earth, then an object that weighs 6 pounds on Earth would weigh about 15 pounds on the giant planet Jupiter.

If available, show *Toys in Space II,* a video from NASA's Liftoff to Learning series.

Introduce the neutral buoyancy pool and parabola flying, two ways in which astronauts experience neutral buoyancy, or weightlessness. As you describe parabola flying, sketch a parabola on the chalkboard to show the path of the airplane. Emphasize the shape of the curve and the directions in which the airplane flies rather than the vocabulary.

Astronauts enter a neutral buoyancy pool, such as the one at NASA's Lyndon B. Johnson Space Center in Houston, Texas. In this pool, called WETF (Weightless Environmental Training Facility), astronauts simulate such outer-space activities as retrieving satellites and repairing instruments like the Hubble Space Telescope. They wear pressurized suits with weights attached to make them neutrally buoyant—they do not sink and cannot swim to the top. As astronauts practice movements in this pool to simulate weightlessness, scuba divers are always present to help the astronauts to the surface in the event of an emergency.

Astronauts participate in parabola flying, as did the actors in the movie *Apollo 13.* On training flights in a former Air Force refueling tanker, NASA crews fly upward at a 45-degree angle to an altitude of about 30 000 feet. Then, following the curve of a parabola, they descend at a 45-degree angle. For about thirty seconds, as the airplane is at the top portion of the curve, the astronauts are weightless.

DEVELOPING THE ACTIVITY

Conduct the following experiments to simulate weightlessness and observe the effects of gravity. As groups of students conduct their experiments, have them watch one another and think about these questions.

- What variables are being changed in the experiments?
- What conclusions can you draw from the experiments?

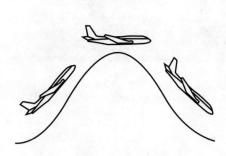

Parabolic path of airplane

Experiment 1: With and Without Weight

Students attach weights to their arms and legs, exercise for five minutes, and record the number of repetitions of each type of exercise. Then students remove the weights, immediately repeat the same exercises, and record the number of repetitions. Finally, have students compare the results and write their conclusions.

Experiment 2: Orbiting Spacecraft

If a lettuce spinner is available, have the class put a few small objects, such as counters or connecting cubes, inside the spinner. Observe as the handle is slowly turned. Gradually increase the spin rate. Ask students to write their observations.

If appropriate, relate the second experiment to students' experiences on carnival or theme-park rides, such as roller coasters. Compare the position of the objects inside the spinner with the position of a Shuttle or satellites in space above Earth. Anything in orbit around Earth has two forces acting on it—forward thrust and gravity. The thrust on the object keeps it moving in the orbit and prevents it from falling to Earth.

After students have shared their observations of these gravity-related experiences, explain that the surface gravity of the Moon and the other planets is not the same as the surface gravity of Earth. Thus, students would not weigh the same on the other planets as they do on Earth.

Ask students if they have seen pictures of how the astronauts who landed on our Moon seemed to bounce easily as they walked. They bounced because the force of gravity is less than on Earth, and they felt lighter. The Moon has a relative surface-gravity value of about 0.17 compared with Earth's value of 1.0.

To develop this concept further for students who are exploring decimal notation, write the names of the planets in alphabetical order with corresponding surface-gravity values in a class chart. These values compare the surface gravity on the planets with the surface gravity of Earth.

If a planet's surface-gravity value is less than Earth's, a student would weigh less than on Earth. If it is greater than Earth's, the student would weigh more than on Earth.

Astronaut working in low gravity on the Moon

Instructional Note: *Base-ten materials will help your students better understand the relative magnitude of decimal numbers. If commercially made flats, longs, and cubes are not available, students can make their own with centimeter-graph paper glued on posterboard. Or, use play money—pennies, dimes, and dollars.*

Ask students to work together in small groups to order the given decimal values. They should use base-ten materials to build models for each decimal number and then match the model with the prepared set of index cards. For example, with a flat as 1, a long as 0.1, and a small cube as 0.01, the model beside Jupiter's index card could comprise two flats, three longs, and four small cubes to represent the value 2.34. Then have students order the models and cards from greatest to least.

Planet	Relative Surface Gravity
Earth	1.00
Jupiter	2.34
Mars	0.38
Mercury	0.38
Neptune	1.19
Pluto	0.06
Saturn	1.06
Uranus	0.92
Venus	0.91

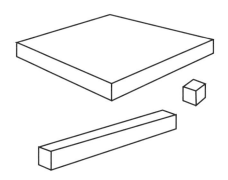

Show students the glyph given on page 84 that represents an 80-pound space creature's weight on Earth. Explain that *glyph* is the name for a figure, or picture, that shows data. For example, if the given picture shows 8 bubbles in the creature's tummy and represents a weight of 80 pounds on Earth, ask what weight is represented by a creature with 8 1/2 bubbles. [85 lb] What does a space creature with 4 bubbles represent?[40 lb] What does it mean if the tummy has more than 3 1/2 bubbles, but fewer than 4 bubbles? [about 38 lb]

Gravity and Weight

Glyph for weight on Earth (80 lb)

Table of Weights

Planet	Weight
Earth	80 lb
Jupiter	187 lb
Mars	30 lb
Mercury	30 lb
Neptune	95 lb
Pluto	5 lb
Saturn	85 lb
Uranus	74 lb
Venus	73 lb

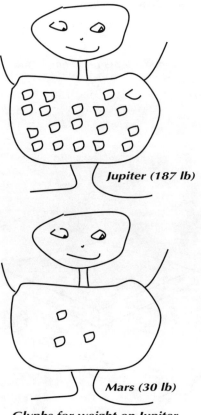

Glyphs for weight on Jupiter and Mars

When students are comfortable using glyphs to represent different weights, ask them to use the values given in the table of weights to make glyphs to represent the creature's weight on the different planets.

CLOSING THE ACTIVITY

Students should share their glyphs and compare them with their other data about the planets. Ask them to describe the patterns and relationships they see. Record students' conjectures, which may include the following:

- The masses of the Moon and the planets are not the same as the mass of Earth.

- Jupiter is the biggest planet and has the greatest surface gravity.

FURTHER EXPLORATIONS

- Students can use calculators to find their weight on the other planets by multiplying their Earth weight by the surface gravity of the planet. For example, a student who weighs 80 pounds on Earth would weigh 2.34 × 80, or about 187, pounds on Jupiter.

- See appendix C for other data about the planets of our solar system. Use these data to provide for students new problems that involve relations among numbers and proportional reasoning. For example, use the period-of-rotation data. Students can then use this information to make up problems for their classmates.

Mission Mathematics: Grades K–6

MISSION TO MARS   K 1 2 3 4 5 6

Students play a probability game to see which spacecraft can reach Mars first. The game models a uniform distribution, so all six spaceships have an equal chance of winning. Students are encouraged to play the game many times to collect data about the number of times each spacecraft wins. They display their data in a bar graph. The more times the game is played, the more likely it is that the students will understand that all spacecraft have an equally likely chance of winning.

PURPOSES

- To play a game that models a uniform probability distribution
- To record and organize cumulative data from repeated probability experiments
- To display and analyze data to identify a pattern
- To identify equally likely events

INTRODUCTION

The United States and Russia have been studying Mars since *Mariner 4* gave us the first close-up pictures of its surface from a flyby in 1965. This initial exploration has been followed by Mariner and Viking missions, which have included both orbiters and landers.

A more recent mission to Mars is the Mars Pathfinder from the United States. Planned future missions include the Mars Global Surveyor from the United States for multispectral mapping, Mars '98 from Russia, a Japanese mission in 1998 to measure solar wind interaction and the upper atmosphere, a European Space Agency InterMarsnet with a network of landers in 2003, and a Sample Return mission by the United States in the early 2000s to retrieve soil samples for analysis.

Why are we studying Mars so intensely? Over the last thirty years, we have learned that Mars is the planet that is most like Earth. It has polar ice caps that have changed over seasons. Evidence of natural canals, or dried riverbeds, indicate that the climate was at one time warmer and wetter. Like Mercury, Venus, and Earth, Mars has gone through volcanism, impact events, and atmospheric changes. Mars has had no water for millions of years and thus has retained more of its surface record of the changes that have occurred. By studying Mars, we can learn more about geologic history than is possible here on Earth.

GETTING STARTED

Explain to students that they are going on a mission to Mars. Since Mars is a long distance away, they need to find which of the six spacecraft will get there the fastest. Review the directions for the game with the class. Students can play as individuals, as partners, or in small groups. Younger students will probably do best with a partner.

Game directions:

1. Roll a number cube (or spin the spinner).
2. On your game-board graph, color the square just above the matching spaceship labeled with that number.
3. Continue in this manner until one column is completely colored. This spacecraft is the winner.

NCTM Mathematics Standards
- Statistics and Probability
- Patterns and Relationships

Materials: Resource page 88, at least six copies for each student; a number cube or spinner with six congruent sections, marked 1–6; crayons or markers; connecting cubes; graphing software (optional)

The planet Mars as seen by the Viking *spacecraft*

Artist's rendering of the Mars Pathfinder

Note to Primary-Grades Teachers: To restrict the data to numbers appropriate for your class, you may want to limit the number of games played by each student or pair of students to two or three.

Technology Connection: *The class data from the table of winners can be graphed on the computer if graphing software is available.*

4. Report your winner to the teacher to be recorded in the class table of winners.
5. Play the game at least six times, making a new game-board graph each time.

Ask students whether they think they will get the same results the next time they play the game. Students should play the game again. Does the same spaceship win this time? Do they want to predict what spaceship will win the third time they play? Students should play the game at least six times and complete a game-board graph each time they play.

DEVELOPING THE ACTIVITY

Students can use connecting cubes to make a concrete graph and model all the data collected. Assign each spaceship a color of connecting cube as well as a number. Students should color their charts using the corresponding spaceship colors.

Each time a game-board graph is completed, students snap together the correct number of cubes for each spacecraft. For example, if in the first game Spaceship 1, the red spaceship, was rolled five times, the student should snap together five red cubes. If Spaceship 2, the yellow spaceship, was rolled eight times, the student should snap together eight yellow cubes. This process is repeated for the other four spaceships, using a different color for each.

As the students complete each new game and game-board graph, they should continue to add connecting cubes to the six sets of cubes. Thus, they are compiling cumulative data for all their games. As the number of games increases, the six sets of cubes will approach equal lengths.

CLOSING THE ACTIVITY

As the students continue playing the game, record the winner of each game for all students in a class table of winners.

Sample Table of Winners

Spacecraft	Number of Wins
Spacecraft 1	ＩＩＩＩ ＩＩＩＩ ＩＩＩＩ ＩＩＩＩ ＩＩＩＩ ＩＩＩＩ ＩＩ
Spacecraft 2	ＩＩＩＩ ＩＩＩＩ ＩＩＩＩ ＩＩＩＩ ＩＩＩＩ ＩＩＩＩ
Spacecraft 3	ＩＩＩＩ ＩＩＩＩ ＩＩＩＩ ＩＩＩＩ ＩＩＩＩ ＩＩＩＩ ＩＩＩ
Spacecraft 4	ＩＩＩＩ ＩＩＩＩ ＩＩＩＩ ＩＩＩＩ ＩＩＩＩ ＩＩＩＩ ＩＩＩＩ Ｉ
Spacecraft 5	ＩＩＩＩ ＩＩＩＩ ＩＩＩＩ ＩＩＩＩ ＩＩＩＩ ＩＩＩＩ
Spacecraft 6	ＩＩＩＩ ＩＩＩＩ ＩＩＩＩ ＩＩＩＩ ＩＩＩＩ ＩＩＩＩ ＩＩＩＩ ＩＩ

Then help the class use the table of winners to make a graph that shows how many times each spacecraft was the winner.

Class Conversation

- Does one spaceship seem to have a better chance of winning than the others?
- Do you think each spaceship has an equally likely chance to win?

Ask students to support their conjectures using the data displayed in the class graph.

FURTHER EXPLORATIONS

- Repeat the activity with a spinner with four congruent sections and a game-board graph with four different spacecraft. Ask students to predict the results of the game.

- Repeat the activity for six spacecraft, but use a spinner that does not have six congruent sections. Ask students to predict whether all spacecraft have an equally likely chance of winning. Then ask the class to play the game with this spinner at least fifty times and record the winners. They should use their class data to decide whether their predictions can be supported with the results of their experiment.

- Students can design new spinners or number cubes and tell whether the game is fair or unfair to all spacecraft.

NCTM Assessment Standards:
Suggestions for assessing students' understanding of this activity include the following:

- *Were students able to record correctly their results on the game-board graphs?*
- *Were students able to interpret the class data to see that there is an equally likely chance for all spacecraft to win the game?*
- *Can students predict whether results will be the same if a different spinner or cube is used?*

Mission to Mars

1. 2. 3. 4. 5. 6.

JOURNEY TO JUPITER   K 1 2 3 4 5 6

This activity should be linked with "Mission to Mars" so students can contrast the mathematical patterns. In both games, students have a similar challenge—they need to travel to a planet by the fastest spaceship available. However, in this game they travel to Jupiter, test only five spaceships, and toss four pennies rather than a number cube.

When a coin is tossed, there are two possible outcomes, heads or tails. The students determine the winning spacecraft by counting and recording the number of heads that appear when they toss four pennies. Students repeat the game several times and record winning spacecraft each time. They analyze their data to find that the probability of zero heads or four heads occurring is far less than the probability of two heads occurring.

NCTM Mathematics Standards
- *Statistics and Probability*
- *Patterns and Relationships*

Materials: *Resource page 93, at least six copies for each student; four pennies; crayons or markers; connecting cubes and graphing software (optional)*

PURPOSES

- To play a game based on a probability experiment
- To record and organize cumulative data from repeatedly playing the game
- To display and analyze data to identify a pattern
- To identify events that are more likely and less likely

INTRODUCTION

Jupiter has been the focus of many of NASA's space-science investigations. *Pioneer 10,* launched in March 1972 by the United States, reached Jupiter in December 1973. From this spacecraft's reports we have learned that Jupiter has intense radiation output, probably has a liquid interior, and has a very strong magnetic field. Later flights of *Pioneer 11* and the two *Voyagers* have taught us about the atmosphere and environment of Jupiter. We have learned that Io, the moon closest to Jupiter, has active volcanism.

In October 1989, the spacecraft *Galileo* was launched from the Space Shuttle *Atlantis*. It passed by Venus and Earth, passed through the asteroid belt, and arrived at Jupiter in December 1995. This mission involves a two-year, ten-orbit probe of the Jovian atmosphere.

The planet Jupiter

GETTING STARTED

After sharing appropriate background information about Jupiter, explain to students that they are going on a journey to that planet. Since it is a very long distance, they need to find which of the five spacecraft will get there the fastest. Review the directions for the game with the class. Students can play as individuals, as partners, or in small groups. Younger students will probably do best with a partner.

Game directions:

1. Place four pennies in a cup. Shake vigorously. Dump the pennies onto a flat surface. Count the number of heads that occur.
2. On your game-board graph, color a square just above the spaceship labeled with the number that matches that number of heads.
3. Continue in this manner until one column is completely colored. This spaceship is the winner.
4. Report your winner to the teacher to be recorded in the class table of winners.
5. Play the game at least six times, making a new game-board graph each time.

Note to Primary-Grades Teachers: *You may want to limit the number of games played by each student or pair of students to two or three.*

Artist's rendering of the Jupiter orbiter Galileo

Demonstrate how to record each throw, or event, on a copy of resource page 93, a game-board graph. For example, when a student throws the pennies and zero heads and four tails show, he or she should color one square for Spaceship 0.

Ask students whether they think they will get the same results each time they play the game.

DEVELOPING THE ACTIVITY

Graph the winning spaceships on a class graph. Then ask questions to help the students interpret their graph.

Class Conversation

- Does one spaceship seem to win more than the others?
- Do some spaceships seem to win a lot? Which ones?
- Do you notice a pattern on the class graph?
- Can you definitely rule out any spaceships as possible choices for your mission to Jupiter? Why?

As the number of games played increases, students should see a normal or bell curve–like pattern emerging on the class graph of winners. Once this shape begins to be noticeable, ask questions such as these:

- Would you choose Spaceship 0 or Spaceship 2? Why?
- Would you choose Spaceship 4? Why?

CLOSING THE ACTIVITY

An important concept to discuss is why no spaceship in the Mission to Mars game seemed to have a clear advantage. In the Journey to Jupiter game, those spaceships with the extreme numbers, 0 and 4, seemed to be at a clear "speed" disadvantage compared with the spaceships with the numbers 1, 2, and 3.

ASSESSING THE ACTIVITY

As students play this game you will have opportunities to observe whether they are able to—

- record their results on a copy of resource page 93, the game-board graph;
- interpret the class graph of winners to see that this game produces a different pattern from the Mission to Mars game;
- make conjectures about why the patterns are different.

FURTHER EXPLORATIONS

Intermediate-grades students may be ready for experiences with tree diagrams to show why the spaceships with the numbers 1, 2, and 3 are more likely to win than the spaceships with the numbers 0 and 4.

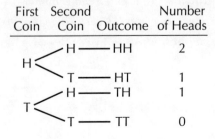

Tree diagram for two coins

Begin with a simpler problem of two coins. Use a tree diagram like the one shown to help students see that four outcomes are possible when two coins are used. Only one outcome is possible with zero heads and only one outcome is possible with two heads. However, there are two outcomes with one head showing, HT and TH. Thus, there are twice as many chances to have an outcome that favors Spaceship 1. The game does not offer an equally likely chance for all spaceships.

The next step is to show a tree diagram for three coins. Ask students to help expand the diagram from two coins to three coins. They will see the eight possible outcomes. Of the eight outcomes, only one outcome gives zero heads and only one outcome gives three heads. However, there are three outcomes with one head showing, HTT, THT, and TTH. There are three outcomes with two heads showing, HHT, HTH, and THH. Thus, there are three times as many chances to have an outcome that favors Spaceship 1 and three times as many chances to have an outcome that favors Spaceship 2. So, the game with three coins heavily favors Spaceships 1 and 2. The game does not offer an equally likely chance for all spaceships.

Finally, ask students to help you extend the tree diagram to show the possible outcomes with four coins. Of the sixteen outcomes, only one outcome gives zero heads and only one outcome gives four heads. However, there are four outcomes with one head showing and four outcomes with three heads showing. Thus, there are four times as many chances to have an outcome that favors Spaceship 1 and four times as many chances to have an outcome that favors Spaceship 3. Six outcomes have two heads showing. Thus, the game with four coins heavily favors Spaceship 2 and favors Spaceships 1 and 3 over Spaceships 0 and 4. The game does not offer an equally likely chance for all spaceships.

First Coin	Second Coin	Third coin	Outcome	Number of Heads
H	H	H	HHH	3
H	H	T	HHT	2
H	T	H	HTH	2
H	T	T	HTT	1
T	H	H	THH	2
T	H	T	THT	1
T	T	H	TTH	1
T	T	T	TTT	0

First Coin	Second Coin	Third Coin	Fourth Coin	Outcome	Number of Heads
H	H	H	H	HHHH	4
H	H	H	T	HHHT	3
H	H	T	H	HHTH	3
H	H	T	T	HHTT	2
H	T	H	H	HTHH	3
H	T	H	T	HTHT	2
H	T	T	H	HTTH	2
H	T	T	T	HTTT	1
T	H	H	H	THHH	3
T	H	H	T	THHT	2
T	H	T	H	THTH	2
T	H	T	T	THTT	1
T	T	H	H	TTHH	2
T	T	H	T	TTHT	1
T	T	T	H	TTTH	1
T	T	T	T	TTTT	0

Journey to Jupiter

RELATED ACTIVITIES IN MISSION MATHEMATICS

The mathematical patterns evident in this game become more obvious when it is played after the game Mission to Mars. You may also want to link this activity with another probability game, Rescue-Mission Game in the "Aeronautics" activities.

Journey to Jupiter

0 1 2 3 4

MISSION TO PLANET EARTH

Children born since humans have gained the ability to take pictures from space have grown up with the image of Earth as a beautiful blue planet that orbits the star we call the Sun. More than earlier generations and because of their interest in the universe, these children have developed a sense that our planet Earth is a closed ecological system. As a system, our environment has a delicate balance that must be maintained to preserve life on Earth as we know it today. Much of this heightened understanding of the delicate environment of our planet has come from information gained through NASA's research. All of NASA's efforts to help us better understand our environment are bundled in an endeavor called "Mission to Planet Earth." Through this research, NASA scientists attempt to answer important questions.

✦ Can we predict catastrophic events related to the movement of Earth's crust?

Earth's crust is constantly moving at about the same rate as our fingernails grow, generally not more than 6 inches a year. This movement contributes to earthquakes, volcano eruptions, and the formation of mountains. A space-age technique for studying this motion is called Satellite Laser Ranging. Two space-age technologies, lasers and satellites, enable us to measure small movements in the crust of Earth. By aiming ground-based lasers at satellites, NASA is gathering data that may someday be important for predicting earthquakes and volcano eruptions.

✦ Can we determine the causes of high-altitude ozone depletion and use this knowledge to reverse this effect?

Ozone is a relatively unstable molecule found in Earth's atmosphere and is crucial to life on Earth. In the upper atmosphere, 15 miles (24 km) above Earth, ozone acts as a shield to protect Earth's surface from the harmful effects of ultraviolet radiation. This ultraviolet shield is important to humans because ultraviolet rays damage the skin in a way that leads to skin cancers. Human activity, especially the introduction of chloroflourocarbons (CFCs) into the atmosphere from aerosol sprays and the combustion of fossil fuels, has had an adverse effect on the ozone layer. NASA satellites, aircraft, balloons, and ground measurements helped detect the effects of CFCs on the ozone layer. From the understanding gained from these observations, we have changed some of our activities. Now we are trying to reduce the production of CFCs worldwide by controling the use of products that produce these harmful effects. Also, we are being more careful of our skin by using lotions that protect it from ultraviolet radiation.

✦ How can we understand Earth on a global scale?

NASA will begin to launch the Earth Observing System (EOS) in 1998. EOS features a series of polar-orbiting and low-inclination satellites for long-term observations of the land surface, biosphere, solid Earth, atmosphere, and oceans. These EOS satellites will be launched over four years. Once in place, the EOS satellites, together with some international Earth-observing satellites, will furnish an Earth-observing capability for at least fifteen years. The data collected will allow scientists to study many aspects of Earth's environment, including the role of clouds, radiation, water vapor, and precipitation; the productivity of the oceans; air-sea exchange; greenhouse gases; changes in land use; the role of polar ice sheets; and the role of volcanoes in climate change.

These and other questions involved in the research associated with Mission to Planet Earth have important implications for future generations. The quality of life on Earth will depend on the ability of scientists to find adequate answers to such questions. The study of Earth using the vantage point of space will be a vital component in the effort to save the environment of our beautiful blue planet.

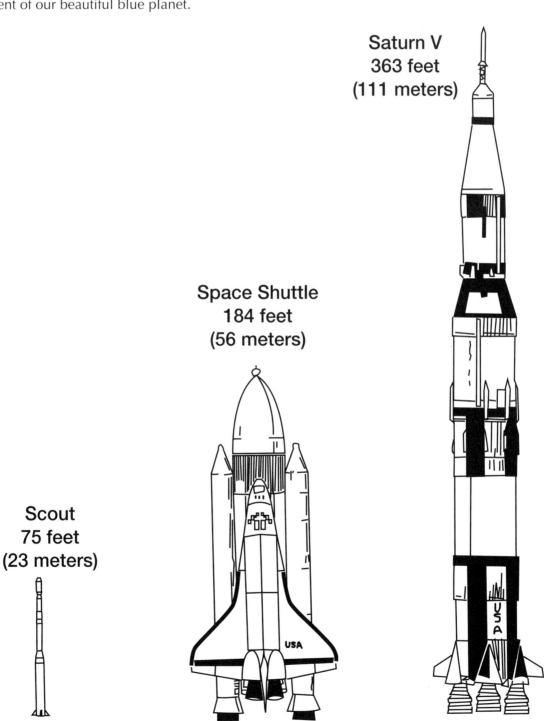

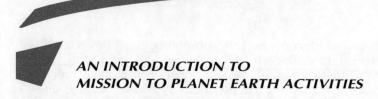

AN INTRODUCTION TO MISSION TO PLANET EARTH ACTIVITIES

NASA's integrated research efforts to study Earth and its changing environment are a perfect centerpiece for interdisciplinary units involving the sciences, geography, international cooperation among governments, commerce and business, communication, and mathematics.

Some of the important ideas that can be explored through the Mission to Planet Earth include natural climate changes, such as volcanic eruptions like Mount Pinatubo in the Philippines in 1991, hurricanes, earthquakes, and El Niño. Another important environmental change is global warming and how it is related to the results of human activities, such as exhaust fumes from cars; smoke from factories; deforestation; and chlorofluorocarbons (CFCs) in refrigeration, air conditioners, and industrial solvents and cleaners.

NASA ocean research has determined ocean heights and circulation patterns and has studied how El Niño affects the rains and floods in some areas, such as California, and droughts in other areas, such as Australia.

The ocean research has also contributed to the development of remote sensing equipment to help commercial fishing operations monitor fish concentrations. Remote sensing data have been used to inventory, map, and assess changes in the quality and quantity of the winter habitats of migratory waterfowl.

NASA and world-health authorities are using remote sensing data to identify, monitor, and model environmental variables that influence mosquito populations and disease transmission.

All activities in this section integrate the four NCTM process standards: problem solving, mathematical reasoning, mathematical communication, and mathematical connections. The following brief outline describes the "Mission to Planet Earth" activities and the associated mathematics.

Activity Name	Mathematics	Grade Levels
Protractor Rocket Launches	Measure distances and angles; analyze data finding range, median, mode, and mean	3–6
Solar Observations Over Time	Elapsed time, patterns and relationships, size and shapes of figures, graphing data	K–6
Collecting the Rays	Linear dimensions, area, volume, time, temperature, geometry in the real world	3–6
Does the Sun Heat Fairly?	Temperature, measuring with Celsius thermometer, choosing an appropriate graph for data	3–6
Weather Watchers	Line graphs, line plots, stem-and-leaf plots; analyzing data	3–6

Other sources of information for Mission to Planet Earth and the related technologies are available through NASA's home page on the Internet. For more information about NASA's materials for teachers, see appendix A.

PROTRACTOR ROCKET LAUNCHES

K 1 2 3 4 5 6

Students compare the flight distances of rubber-band rockets when the launch angle and the amount of force vary.

PURPOSES

- To measure angles with a protractor
- To measure distances
- To collect, organize, analyze, and interpret data
- To find the range, median, and mean of a set of data
- To use estimation to determine the reasonableness of sums found with a calculator

INTRODUCTION

Much of the data collected by NASA about our Earth and its atmosphere has been obtained with unmanned small satellite missions, high-altitude probes, and reentry experiments. Each year, NASA launches an average of thirty sounding rockets. The name derives from the nautical term *to sound,* or to take measurements. Fourteen different sounding rockets come in sizes that range from 7 feet to 65 feet.

Sounding rockets fly for fewer than 30 minutes, can reach altitudes of more than 400 miles, and carry payloads of instruments for experiments. The payload reenters the atmosphere, falls to Earth with a deployed parachute, and is recovered. These inexpensive rockets allow scientists to conduct investigations about the upper atmosphere, the Sun, stars, galaxies, and other planets.

Over the years, NASA has used eight major space vehicles that range in size from the 75-foot Scout to the 363-foot Saturn V rocket to perform suborbital tasks as well as to launch interplanetary exploration. Unmanned Scout launch vehicles have been successfully used to place payloads into orbits at much less expense than Shuttle missions. The Saturn V is a better known rocket. It was used for the Apollo missions and for placing the *Skylab Space Station* into Earth orbit. Since 1981, the most familiar rockets are those used for the Space Shuttle.

Invite your students to explore variables involved with rocket launches—the amount of thrust and the angle of launch. They can pretend to be NASA scientists who conduct experiments; collect, analyze, and interpret data; and make conclusions.

GETTING STARTED

Students explore differences in the distances traveled by rubber bands launched from various angles. The three angle measurements to be tested are 30 degrees, 60 degrees, and 90 degrees. The size of the rubber bands tested remains constant, such as a circumference of 10 cm and a diameter of 3 cm. In the first experiment, the amount

NCTM Mathematics Standards
- *Measurement*
- *Statistics*
- *Patterns and Relationships*
- *Number Sense and Numeration*
- *Estimation*

Materials: *Three protractors, three rulers, tape, rubber bands of the same size, metersticks or centimeter tape measures, clay (optional)*

**Saturn V
363 feet
(111 meters)**

**Space Shuttle
184 feet
(56 meters)**

**Scout
75 feet
(23 meters)**

Safety Notes: *Have students wear goggles during the launches. Do not allow students in the landing area while the launches are being conducted!*

Management Tip: *You may want small groups of students to work together as launch teams. Each team should experience at least two or three launches with each of the three launch angles.*

of stretch in the rubber band also remains constant because each rubber band is pulled to the same 7-cm mark on the ruler of the launch pad.

If this is your students' first experience with protractors, you will want to take some time to introduce this new tool for measuring angles. Let the students explore how the ruler is tilted differently when they place it to form the 30-degree, 60-degree, and 90-degree angles.

Share the following directions with your students.

- Prepare three launch pads. First, securely tape a ruler to a protractor, centimeter side facing up, to form a 30-degree angle. In taping the rulers to the protractors be sure not to let the ruler extend below the straightedge base of the protractor. The protractors are set upright on a flat surface for the rubber-band launches and, therefore, need a level bottom surface. Next, tape a second ruler to the second protractor to form a 60-degree angle. Finally, tape the third ruler to the last protractor to form a 90-degree angle.

- Make a table on a data-collection sheet. You need one column to identify the launch and three columns to identify the angle of the launch pad. Mark the first column "Launch"; the second column, "30 degrees"; the third column, "60 degrees"; and the last column, "90 degrees." You need one row in your table for each launch.

- Prepare the launch site by placing at least six metersticks end to end or by laying out six meters of measuring tape. To make a more stable launch pad, put some clay on each end of the protractors and anchor them to the floor or ground at the front edge of the first meterstick.

- Launch your rubber band by looping it over the free end of the ruler, pulling it back until it reaches the 7-cm mark, and letting go.

DEVELOPING THE ACTIVITY

Ask students to predict the results of the launches.

Class Conversation

- Which launch pad do you predict will produce the longest flights?
- Which launch pad will produce the highest flights?
- Which launch pad looks most like the one from which the Space Shuttle is launched, in terms of the launch angle?
- Where do you think the rubber bands launched at the 60-degree angle, the middle angle, will fall? Will their distances be exactly half-way between the distances for the rubber bands launched at the 30-degree and 90-degree angles?

Students should now begin launching their rubber-band rockets and collecting data. The sample data on page 99 were gathered by using rubber bands with a circumference of 10 cm and a diameter of 3 cm. The rubber bands were stretched to 7 cm on the ruler. It is important to use rubber bands of the same size and to be very careful in letting go of the rubber bands at exactly the same place on the ruler each time—in this example, from the 7-cm mark.

DISPLAYING AND ANALYZING THE DATA

Ask students to look at their data and make conclusions about their experiment. Their first observations will probably be that the rockets launched at the 30-degree angle traveled the longest distances and those launched at the 90-degree angle traveled the shortest distances.

Then ask students to think about how they can describe the set of data for each launch angle.

Sample Flight Distances
(in centimeters)

Launch Number	30°	60°	90°
Launch 1	416	300	5
Launch 2	433	302	20
Launch 3	447	330	25
Launch 4	448	330	29
Launch 5	450	332	30
Launch 6	452	338	40
Launch 7	463	350	46
Launch 8	467	366	58
Launch 9	468	368	63
Launch 10	468	374	66

Class Conversation

- How can you summarize the distances of the ten launches for the 30-degree-angle launches?
- Can you use just one number to tell about how far the rubber-band rockets traveled?

Introduce or review the concepts of median and mean. To find the median of the data, students must order the data from least to greatest. They can make a line plot or stem-and-leaf plot to do this. Either display provides an organization that enables students to recognize easily the least and greatest values and then to compute the difference of these two values to find the range of the data.

Management Tip: *You can combine the data collected by all the launch teams to make a large set of data for the whole class to analyze.*

Students should use calculators to calculate the mean distance for each of the three angles. Ask students to estimate the sum of the ten distances in each column before they add. One way to estimate is to choose a middle value in the line plot or stem-and-leaf plot and multiply the value by 10. For example, they can find 10 × 450 = 4500 for the estimate of the sum of the ten distances flown from the 30-degree launch angle. This process will help them know whether the sum showing in the display of the calculator is reasonable.

Data Analysis of Sample Flight Distances
(in centimeters)

	30°	60°	90°
Range	52	74	61
Mean	451.2	339	38.2
Median	451	335	35

CLOSING THE ACTIVITY

Encourage the class to discuss the findings of their experiment, using the range, mean, and median. In the sample data, the mean and median were very similar to each other in each of the three data sets. The distances for the 90-degree-angle launches had a greater range than the flight distances for either of the other angles.

After students have analyzed and interpreted the data for the three launch angles, ask them to share their conclusions with the class. You may want them to work together in groups to write a report about their experiment.

NCTM Assessment Standards:
Assessment should be a means of fostering growth toward high expectations. Encourage students to include the following in their reports:

- *A description of the purposes of the experiment*
- *Their predictions*
- *Their data log*
- *A display that shows the data distribution*
- *Results of their data analysis*
- *Their interpretation of the data analysis*
- *Their conclusions*

FURTHER EXPLORATIONS

- Conduct the investigation again but this time increase the stretch, or force, on the rubber band. Make predictions about how this change in force will affect the resulting distances before gathering the new data. Collect data at several points: 7-cm stretch, 10-cm stretch, 15-cm stretch, and so on. Does a pattern emerge? What is the effect of an increase in force on the distance traveled?

- Conduct the investigation again, this time varying the launch angles by using such angles as 15 degrees, 45 degrees, and 80 degrees. Collect data using the same size of rubber bands and the same amounts of force, or stretch, as in the earlier investigations. How do these results compare with your original data? What patterns do you see? What is the relationship between the launch angle and the flight distance?

- Use decimals to record flight distances. For example, 345 centimeters would be recorded as 3.45 meters.

RELATED ACTIVITIES IN MISSION MATHEMATICS

Students make line plots and stem-and-leaf displays of data in the activities "Long-Distance Airplanes," "Target Practice," and "Weather Watchers." Students also explore rockets in "Fizzy-Tablet Rockets."

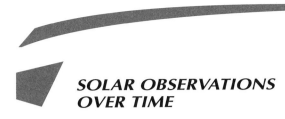

SOLAR OBSERVATIONS OVER TIME   K 1 2 3 4 5 6

My Shadow

I have a little shadow that goes in and out with me,
And what can be the use of him is more than I can see.
He is very, very like me from the heels up to the head;
And I see him jump before me, when I jump into my bed.

The funniest thing about him is the way he likes to grow—
Not at all like proper children, which is always very slow;
For he sometimes shoots up taller like an india-rubber ball,
And he sometimes goes so little that there's none of him at all.

He hasn't got a notion of how children ought to play,
And can only make a fool of me in every sort of way.
He stays so close behind me, he's a coward you can see;
I'd think shame to stick to nursie as that shadow sticks to me!

One morning, very early, before the sun was up,
I rose and found the shining dew on every buttercup;
But my lazy little shadow, like an arrant sleepy-head,
Had stayed at home behind me and was fast asleep in bed.

—Robert Louis Stevenson

NCTM Mathematics Standards
- *Geometry and Spatial Sense*
- *Statistics*
- *Patterns and Relationships*

Materials: *The poem "My Shadow" by Robert Louis Stevenson; 6" × 9" rectangle of oak tag; stapler or hammer and thumbtack or small nail; Old Farmer's Almanac or other publication listing the time of local sunrise and sunset; inch or centimeter graph paper; small mirror made of plastic or metal; sticky-backed dots (3/4" or 1" in diameter); globe; lamp with bare lightbulb; large chart paper; 4" × 6" index cards (one for each pair); yardstick; yellow and black crayons; chalk; step ladder (optional)*

Students develop an understanding of Earth's relationship to the Sun by observing shadows and by graphing the number of hours of daylight during the school year.

PURPOSES

- To observe how the shadow of an object changes during the day
- To gather and record data over a long time
- To observe that the number of daylight hours changes over the year
- To display data on a graph
- To read a timetable
- To determine elapsed time

INTRODUCTION

NASA scientists continue to study the motion and locations of Earth and the other planets of our solar system. They collect data using the equipment and instruments on spacecraft and on satellites. Mission to Planet Earth is a 15-year project to collect data so we can understand the processes and patterns of the total Earth system.

Invite your students to pretend that they are NASA scientists who are to collect data about the movement of Earth around the Sun.

BEFORE THE ACTIVITIES

From this set of three activities, choose those appropriate for your class.

Before your students begin the activities, you may want to use the following model to review or introduce revolution and rotation.

Place a bare lightbulb in the middle of the darkened classroom and turn it on to represent our Sun. One student can hold the globe at about a 23-degree angle to the Sun and walk around the Sun in an oval, or egg-shaped, path as a simulation of Earth's path or orbit. Another student should walk along and rotate the globe at the same time to show the change from day to night to day.

Background Information: *As appropriate for your students, you can share the following information:*

- *Earth's axis is at a 23° 27' angle to the Sun.*
- *Earth rotates once every 23 hours 56 minutes, but our clocks are set to a 24-hour day.*

101

Stop the students when the North Pole points toward the Sun. Ask what season this position represents for us. Ask questions to help them see that in the Northern Hemisphere, the Sun is visible for a longer time, and, therefore, the days are longer. Help them recall their own experiences of summer days that are very long. Discuss some of the northernmost locations that have days when the Sun never sets. Then discuss what is happening in the southernmost latitudes.

Have the students continue walking around the Sun. When the position is reached that represents winter, again examine the amount of daylight in the Northern Hemisphere and the angle of the Sun, and compare what happens to the Northern and Southern Hemispheres. Continue the experiment until all four seasons have been discussed and compared with one another.

Part 1: Shadows and the Sun
GETTING STARTED

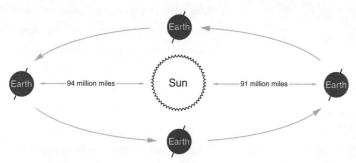

This activity should be started at the beginning of the morning on a sunny day. After reading "My Shadow" by Robert Louis Stevenson, hold a 6-inch-by-9-inch oak-tag rectangle in front of the class. Ask the students to describe what they think the shadow of the rectangle will look like.

Take students outside and hold up the rectangle so that the shadow on the ground does not look like the oak-tag rectangle with respect to either size or shape. Have one student trace the shadow on the pavement or on a piece of paper. Compare the tracing with the original.

NCTM Assessment Standards: *Some assessment suggestions that can be integrated into the activity include these:*

- *Ask students to draw pictures of themselves that include the Sun and their shadows in early morning and at noon.*
- *Ask them to think about the lengths of their shadows compared with their heights and their position relative to the Sun and their shadows. The shadow should be longer than the height of the student in the early morning and shorter than the student at noon. The student's body should be between the Sun and the shadow in both pictures. Some children will position the Sun lower in their morning picture than in their noon picture.*

Class Conversation

- Is the tracing the same size or shape as the original figure? What differences do you see?
- Why do you think this happened?
- How can we move the rectangle so that the shadow looks just like the original?

Try the students' conjectures by moving the position of the rectangle and tracing the shadow again. Distribute the 4-inch-by-6-inch index cards to pairs of students and let them experiment with shadow sizes and shapes. Have students discuss their results.

DEVELOPING THE ACTIVITY

Mount the 6-inch-by-9-inch oak-tag rectangle on the end of the yardstick. Plant the yardstick perpendicular to the ground with the rectangle facing due south. Place a large piece of paper on the shadow and trace it. Secure the paper so that it will not be moved during the school day.

Class Conversation

- What do you think the shadow will look like in an hour?
- Where will the shadow be?

After discussing their predictions, take the class inside. Return in an hour and trace the shadow again.

- Do you see any changes in the shadow? [Elicit responses that include size, shape, and location.]
- Why do you think these changes happened?
- When we return in an hour, what do you think you will see?

As schedules allow, return every hour during the day. Be sure that students make an observation at, or close to, noon. Ask the class for their predictions after each observation. Students may be surprised that the shadow will get bigger after noon.

CLOSING THE ACTIVITY

Near the end of the school day, look at the pattern of the tracings of the shadows.

Class Conversation

- Why do you think the shadow moved? [The position of Earth moves in relation to the Sun.]

- Why did the shadow change in size? [As the Sun appears to be higher above the horizon, the angle is greater.]

- What happens to your shadow at different times during the day? [It changes from longer to shorter to longer.]

Part 2: Angles of the Sun

GETTING STARTED

Early in the morning of a sunny school day, take a pocket mirror to the window. Catch the Sun's rays with the mirror and reflect them around the room. Play with the mirror by reflecting the rays onto the walls, ceiling, and floor. Let students move the mirror to see where they can make the rays go. Have students describe how the reflection moves as they move the mirror.

Find a spot on the windowsill or window frame where the mirror can be permanently mounted so the reflection will be on the ceiling or high on a wall. (This activity requires windows that are not facing north.) Mount the mirror and watch the movement of the reflection during the day. Record the time of each observation on a sticky-backed dot, and place the dot in the middle of the reflection.

Class Conversation

- What do you think will happen to the reflection during the day? Will it stay in the same place?

After one hour, have students observe the location of the middle of the reflection and place there a new dot marked with the time.

- Where is the reflection now? How far has it moved?

- Where do you predict it will be in an hour? This afternoon? Tomorrow morning?

- Why is the reflection moving even though the mirror is staying in the same position?

Continue observing, marking, and making predictions each hour throughout the day. Near the end of the school day, have the students summarize their observations.

DEVELOPING THE ACTIVITY

On the next sunny day, review the activity. Take down the sticky-backed dots. Select a time during the day when the Sun shines on the classroom window, for example, 11:15 A.M., and put up just one dot. Tell students that they will mark the reflection at this same time on as many days as they can during the school year. Ask why they might not be able to mark every day. [rain, holidays, and so on]

Each day, record the date on a sticky-backed dot and stick it to the middle of the reflection.

Note to Intermediate-Grades Teachers: *During the activity, if students trace the shadow on grid paper, they will find a close estimate of the area of the shadow. They can then graph the areas of the shadow as it changes over time. When is it greatest? When is it the least? Is there a pattern? Why does it change shape?*

NCTM Assessment Standards: *Observe students to see their willingness to observe, detect, and predict changes in the location of the reflection. Some students may conclude that the location of the reflection changes as the angle of the Sun changes.*

Solar Observations Over Time

Note to Intermediate-Grades Teachers: *Compare the angles over the course of two months by running a string from the mirror to the center of the reflection. Measure the angles with a protractor and record the measures. Is there a relationship between the facts that the days get shorter in the winter and the angle of the Sun to Earth is lower?*

CLOSING THE ACTIVITY

Throughout the year, mark the reflection at the same time of day. After a few weeks, ask what pattern the class can observe and what conjectures they have about why this movement occurs.

By holding a string from the mirror to the dot, you can see the angle of reflection. Discuss how the changes of the angle of the Sun to the mirror is related to the angle of reflection.

Make regular observations during the year. Carefully note the movement as seasonal changes occur, such as when the days change from getting shorter to getting longer at the end of December and throughout January.

Part 3: Length of Daylight
GETTING STARTED

Discuss the cycle of changing periods of daylight during the year.

- Did you watch fireworks last summer? When did you see them? Why were they set off so late at night? (Many students will not remember that it does not get dark enough for fireworks until after bedtime in the summer.)
- About what time does it get dark now?
- About what time does it get dark on New Year's Day?

DEVELOPING THE ACTIVITY

Ask students to estimate how many hours of daylight are in a day. After taking estimates, tell your class that they will be graphing the number of daylight hours in each day during the year. Talk with them about the length of each day being 24 hours.

Note to Intermediate-Grades Teachers: *During the first two weeks, help students look up the times and make the rectangle for the graph as a whole-class activity. After the students know the procedure, assign a pair of students to do this task daily. Change pairs every week throughout the year.*

Use the almanac to get the time of sunrise and sunset at your geographic location. When recording, round times to the nearest half or quarter hour, as appropriate for your students.

To make the class graph on a large piece of chart paper, turn the chart paper so the longer dimension is used for the horizontal axis. Draw horizontal and vertical axes so they intersect at right angles at the left side of the paper.

Using centimeter grid paper, cut a 1 centimeter × 24 centimeter rectangle for each day, and paste it on the class graph above the horizontal axis. Place the 24-centimeter dimension vertically. Each square in the rectangle represents 1 hour during a day. Label the top and bottom lines "12:00 midnight" and the middle line "12:00 noon." Label the date on the horizontal axis.

Note to Primary-Grades Teachers: *The increments in the difference in daylight from one day to the next may be too small for younger students to understand. Make the graph with 1-inch grid paper. Record the times of sunrise and sunset on a selected day once each week. Ask students to tell whether the times are staying the same each week or are changing. After several weeks, ask what patterns they see.*

Help the students determine the elapsed time between sunrise and sunset. Then locate and color yellow the daylight hours on the rectangle, and color black the night times. The ends of each rectangle should be black and the middle portion should be yellow.

After the first day is recorded and pasted into place, have students discuss and record appropriate labels for the vertical and horizontal axes and compose a title for their graph.

During the first two weeks, very little change from day to day will be apparent. However, as the graph continues, the small daily changes will become more noticeable to the class. As needed during the year, add a new sheet of chart paper to extend the graph.

CLOSING THE ACTIVITY

By the end of the year, the graph will be almost 2 meters long! If you keep the graph rolled up, every few weeks unroll the graph and study the patterns. Then have students make predictions of future patterns.

As the graph is constructed over the school year, students should be able to observe the changes that occur across the seasons. They should conclude that the daylight hours are longer during the summer months and shorter during the winter months. They may be able to use recorded data to predict the approximate time that it will get dark on a given day.

NCTM Assessment Standards: *It is very important to watch for each small step that students make in learning to communicate their reasoning to others and to praise them for their progress.*

Solar Observations Over Time

COLLECTING THE RAYS

*K 1 2 **3** 4 5 6*

Students explore how variations in solar collectors affect the energy absorbed. They make rectangular prisms that have the same volume but different linear dimensions. Students investigate relationships among the linear dimensions, the area, and the volume of rectangular prisms.

NCTM Mathematics Standards
- *Geometry and Spatial Sense*
- *Measurement*
- *Estimation*
- *Statistics*
- *Patterns and Relationships*
- *Number Sense and Numeration*

PURPOSES
- To develop spatial sense
- To explore linear dimensions, area, volume, time, and temperature
- To relate geometric ideas to number and measurement ideas
- To recognize and appreciate geometry in the world

Materials: *For each group: resource page 109; oak tag or lightweight cardboard; masking tape; scissors; rulers; 1" × 4" × 6" oak-tag box constructed by the teacher; thermometers; thirty-six 1" cubes*

INTRODUCTION

The power used by satellites and by astronauts living in a space station comes from the Sun. Energy is collected with solar panels and converted into electricity. Here on Earth, we use passive solar collectors to capture the Sun's heat energy.

BEFORE THE ACTIVITY

Use the pattern on resource page 109 to make a 1 inch × 4 inch × 6 inch open, oak-tag box for each group.

International Space Station

GETTING STARTED

Discuss the solar panels on the space station. Brainstorm with students and then list the features that they think are necessary for a good solar collector. Will shape make a difference?

After students have had time to discuss different ideas, suggest that they conduct experiments to explore the effects of changing the linear dimensions of one possible shape for a solar collector, a rectangular prism.

Ask students to plan how to construct several boxes with the same volume but with different measurements for the length, width, and height.

Show students a preconstructed box and the 1-inch cubes. Ask them to work with their group to estimate how many cubes will fill the box. Each group should be prepared to support its estimate.

DEVELOPING THE ACTIVITY

Ask a representative from each group to give its estimate and explain its reasoning. Then have students find the volume of the box using the 1-inch cubes. They should find that twenty-four 1-inch cubes fit into the box, so the volume of the box is 24 cubic inches.

Technology Tip: *Providing calculators for students to use during the factoring activity may facilitate reasoning for students who have not yet mastered multiplication facts.*

Ask students to use the cubes to find other rectangular prisms that have a volume of 24 cubic inches. Show how to record the factor triple for the 1 inch × 4 inch × 6 inch box as (1, 4, 6). Discuss the order of the dimensions in each triple. Show which dimension corresponds to the 1-inch measure, the 4-inch measure, and the 6-inch measure.

Ask students to work in groups to find other factor triples of 24. They can use the blocks to model the rectangular prism that corresponds to each triple.

Record on a class chart the factor triples for the different prisms that the class discovers.

Ask students to explore other ways to determine the volume of boxes without using the cubes. Let them work with their groups to discover the formula for the volume of a rectangular prism, $l \times w \times h$.

After they have discovered the formula, ask students to find the factor triples that correspond to a volume of 36 cubic inches. Assign each group the task of constructing one of the rectangular prisms with a volume of 36 cubic inches. Each group should be prepared to prove to the class that its rectangular prism has a volume of 36 cubic inches, either by applying the formula or by filling the prism with cubes.

Some students may decide to use the formula to explore sides with decimal or fraction dimensions, such as 0.5 inch $\times$ 9 inches $\times$ 8 inches.

When constructing the boxes, students may need to add interior supports. To have consistent measures in the following experiment, be sure that the supports do not restrict the flow of air.

After the students have constructed their boxes, ask this question:

- Which of these boxes with a volume of 36 cubic inches will collect the most solar energy?

With the class, discuss procedures for the experiment. Encourage their decisions about such factors as these:

- Should the boxes be located indoors or outdoors?
- How and where should the thermometers be placed in the boxes? On the floor of the box? With the bulb in the box? With the thermometer taped to the back of the box so that the bulb is inside the box and the tube outside? Should they cover the box with clear material and read through the "window"?
- What intervals of time should be between temperature readings? Fifteen minutes? An hour?

Make sure that all boxes are the same color and have no insulation.

Each group of students should check on its own box. Each student in the group should independently collect, organize, and display the data on an appropriate graph.

CLOSING THE ACTIVITY

After the experiments are complete, have the students post their data and graphs next to the boxes. As a whole class, discuss the results.

Class Conversation

- Which boxes got the hottest? Why?
- Did some boxes heat faster? Why?

Note: The most dramatic findings will occur outside on a cool, windless day.

FURTHER EXPLORATIONS

- Ask questions such as the following to help students think of ways they can extend this activity.

Whole-Number Factor Triples of 24

(1, 1, 24)　(1, 2, 12)　(1, 3, 8)

(1, 4, 6)　(2, 2, 6)　(2, 3, 4)

Instructional Notes: *Students may have difficulty understanding that boxes with different dimensions may have the same volume. Adult helpers may need to help students construct the boxes to conduct the solar exploration and investigation.*

Whole-Number Factor Triples of 36

(1, 1, 36)　(1, 2, 18)

　　　　(1, 3, 12)　(1, 4, 9)

(1, 6, 6)　(2, 2, 9)

　　　　(2, 3, 6)　(3, 3, 4)

Algebraic Thinking: *As students explore the formula for the volume of a rectangular prism, they will have the opportunity to explore the commutative property of multiplication—the product is the same regardless of the order of the three factors.*

Instructional Note: *If available, encourage students to use graphing software as they collect and organize their data.*

Collecting the Rays

NCTM Assessment Standards: *Ask students to respond to the following in their journals: Suppose that you are a NASA scientist and must write a report about the findings of your experiments. Work with your group to write a report. Be sure to include the purpose of the experiment, the procedures you used, the data you collected, your findings, and recommendations of which dimensions resulted in the best solar collectors.*

—How is the area of the top of the box related to the efficiency of the solar collector?

—Do you think that the color of the box can make a difference?

—What other factors might make a difference?

—Suppose that we have a race to reach 100 degrees Fahrenheit fastest. How would you design your box for the race?

—What is the greatest difference between the temperature of the air outside a box and inside a box that your class can get?

—How would insulation in a box affect your solar collector?

- Students may want to investigate the use of solar cells in homes and businesses. Is solar energy a viable energy source? What about other forms of energy sources? What are alternatives to fossil fuels?

- Several kits, such as Solar House, Solar Oven, and Solar Energy, are available from technology and science suppliers that deal with solar power and energy.

RELATED ACTIVITIES IN MISSION MATHEMATICS

In the activity "Weather Watchers," students collect and organize their own weather-related data. They also choose an appropriate graph for displaying their data. In the activity "Does the Sun Heat Fairly?" students explore how different materials on Earth absorb and reflect the Sun's energy.

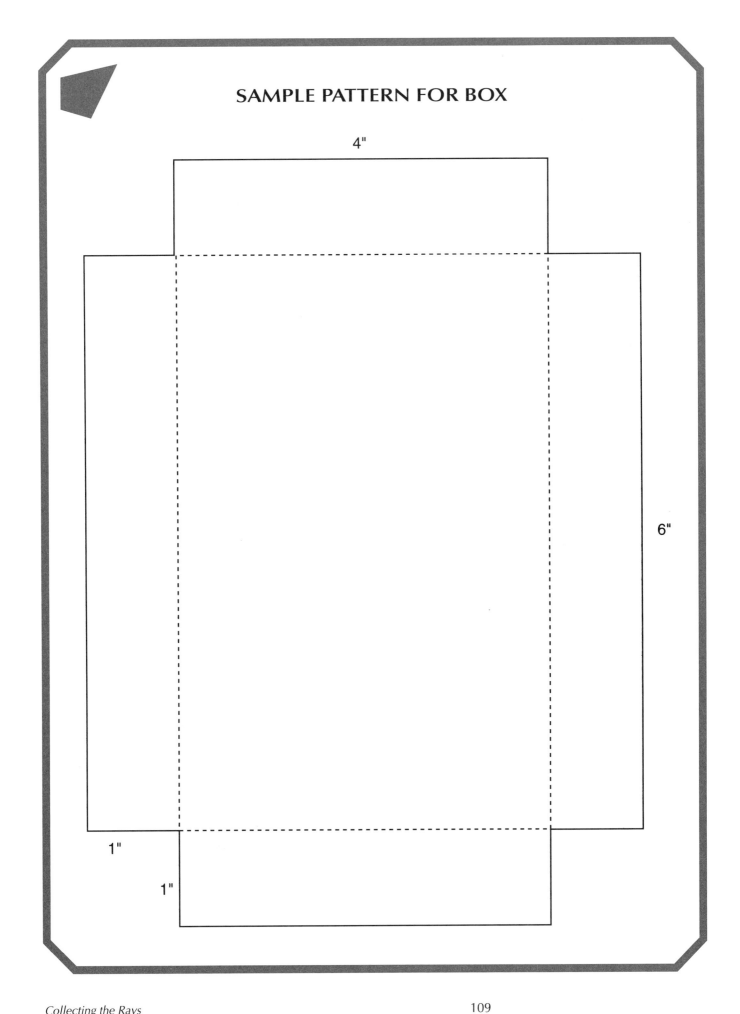

SAMPLE PATTERN FOR BOX

Collecting the Rays

DOES THE SUN HEAT FAIRLY?   K 1 2 3 4 5 6

Students determine which Earth material—pebbles, white sand, or black dirt—absorbs the most heat from the Sun by making conjectures and observations and using their reasoning skills to analyze data.

NCTM Mathematics Standards
- *Measurement*
- *Statistics*
- *Patterns and Relationships*
- *Number Sense and Numeration*

Materials: *For each group: three Celsius thermometers; three clear plastic cups; pebbles, white sand, and black dirt; resource page 112*

PURPOSES

- To build an understanding of temperature
- To experience the process of measuring with a Celsius thermometer and to develop concepts related to units of measurement
- To make and use measurements to solve problems
- To collect data over time
- To organize and describe data
- To construct, read, and interpret displays of data

INTRODUCTION

As part of Mission to Planet Earth, a multiyear, global research program, NASA has been able to make images of the global biosphere of plant life both on the land and in the oceans by using data from different satellites. Ocean images of plant life were provided by the *NIMBUS-7* satellite, launched in October 1978. Land images were from a satellite launched in June 1981, the *Advanced Very High Resolution Radiometer*. During 15 000 orbits of Earth, this satellite measured land-surface radiation, which is a measure of the potential for vegetation production.

Understanding that different materials on Earth absorb the Sun's radiation at different rates can help your students better understand why NASA's satellite images of the oceans and land surfaces show different patterns.

GETTING STARTED

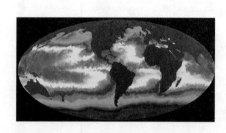

Pose this question to the students: Which of these materials—pebbles, white sand, or black dirt—will absorb the most heat from the Sun?

Let the students volunteer their predictions. Ask them to give reasons for their answers, which should be recorded on chart paper or on an overhead transparency.

Ask the students how they can test their predictions, or hypotheses. Students can discuss this question in their groups and describe their proposed methods. If their methods do not include opportunities to gather data to answer the question, suggest the following activity.

Have the groups fill each container with an equal amount of one of the three materials. Give each group a copy of resource page 112, a data-collection chart.

Students then place the three containers outside in the sunlight. Students should read and record the temperature of the materials at the start of the experiment and after 1/2 hour, 1 hour, 1 1/2 hours, 2 hours, and 2 1/2 hours. Students should record their findings on their group charts.

Class Conversation

- Looking at the data in your chart, can you draw any conclusions? What makes you think that?
- Does it look like one material absorbed more of the Sun's energy faster than the others?

- Can you think of a way in which we can display our data to show patterns that can help us draw conclusions?

DEVELOPING THE ACTIVITY

Encourage students to discuss their ideas for displaying the data. They may suggest many different graphing ideas, such as line and bar graphs, stem-and-leaf plots, or pictographs. Ask, Which is the best way for our data? How can we decide?

Since these data record changes over time, you may want to suggest that students use a line graph. Some students may need to make a separate line graph for the temperatures recorded for each soil before they are ready to make one multiple-line graph showing the data for all three on the same graph. Let each group graph its results.

Remind students that a graph needs to be complete so that everyone who reads it can understand what it shows about the experiment. Have them label the horizontal and vertical axes and compose an appropriate title.

Class Conversation

- On the line graph, how can we draw the lines to help us know which represents the data for the different types of soil? (You may need to suggest dashed or broken lines or different colored lines for children who do not have their own suggestions.)
- To help students remember that they need a key or legend to explain how the different sets of data are represented, ask, How can we make sure that everyone who reads our graphs will know which lines represent the pebbles, dirt, and sand?

CLOSING THE ACTIVITY

As students analyze their data using their graphs, help them interpret their data with the following generalization: Lighter colors reflect more of the Sun's energy; darker colors absorb more and, therefore, get warmer.

Because different Earth materials absorb and reflect the Sun's energy at different rates, weather and wind cycles are created and affected differently.

Meteorologists use NASA satellites and ground measurements to collect data that they can use to predict the weather.

Class Conversation

- Why do we use graphs to show our data?
- How do graphs make it easier to understand our data?
- How do these graphs help us communicate our information and ideas to others?

Students should attempt to draw conclusions from their graphs. Ask students to record in their journals their reflections about their experiences and findings.

FURTHER EXPLORATIONS

Students may want to make conjectures about and then repeat the experiment with other materials and with water.

If available, students can graph their data using computer software programs.

RELATED ACTIVITIES IN MISSION MATHEMATICS

In the activity "Catching the Rays," students explore prisms with the same volume but different dimensions that can be used as solar collectors.

NCTM Teaching Standards: *Good mathematical tasks do not separate mathematical thinking from mathematical concepts or skills. The tasks capture students' curiosity and invite them to speculate and to pursue their hunches.*

In this activity, students construct number meanings through real-world experiences and the use of physical materials; they interpret the multiple uses of numbers encountered in the real world.

NCTM Assessment Standards: *Assessment opportunities should be planned to help your students become independent learners. You may want to discuss with students how they can become involved in self-assessment as they repeat the activity. Help them think of criteria for assessing their data-collection, graphing, and interpretation skills and their process skills, such as reading a thermometer.*

Does the Sun Heat Fairly?

DOES THE SUN HEAT FAIRLY?

Problem: Which material heats up the fastest in the Sun—pebbles, white sand, or black dirt?

Directions

1. Fill each container with an equal amount of one of the three materials.

2. Read and record the temperature of each material when you start your experiment and then every half hour until you have completed this table.

Data-Collection Table

Temperatures (in degrees Celsius)

Materials	Time					
	Start	$\frac{1}{2}$ hour	1 hour	$1\frac{1}{2}$ hours	2 hours	$2\frac{1}{2}$ hours
Pebbles						
White sand						
Black dirt						

WEATHER WATCHERS   K 1 2 3 4 5 6

Students collect and analyze data about the weather and learn to make a stem-and-leaf plot. Students need access to a newspaper or other sources for collecting weather data.

PURPOSES

- To collect, organize, and describe data
- To construct, read, and interpret different displays of data—line graphs, line plots, and stem-and-leaf plots
- To formulate and solve problems that involve collecting and analyzing data

INTRODUCTION

Mission to Planet Earth is a program of major importance at NASA. Its purpose is to study scientifically whether Earth's climate is changing and what the positive and negative contributions of human activities might be.

Scientists use instruments in satellites to measure the interactions of Earth's land, oceans, and atmosphere. Data from the satellites are transmitted as electronic signals and are changed to measurements that are useful for studying weather. For example, the data can be used to make maps of Earth's cloud cover.

NASA scientists also gather and use data to discover possible long- and short-term climate changes. By taking measurements in space and on the ground, they can decide if a connection exists between what is happening in the atmosphere and in space.

NASA and the National Oceanic and Atmospheric Administration (NOAA) supply local news programs with information for the weather reports. Managed by the NASA Goddard Space Flight Center in Greenbelt, Maryland, the Earth Observing System (EOS) includes a series of polar-orbiting and low-inclination satellites that observe and record information about the land surface, biosphere, solid Earth, atmosphere, and oceans.

Invite your students to pretend that they are NASA scientists collecting data to help them predict the weather.

BEFORE THE ACTIVITY

You may want to invite a local meteorologist from a radio or television station to visit your class. Often they have materials and suggestions for other age-appropriate activities. There is usually no charge for these visits.

NCTM Mathematics Standards
- *Measurement*
- *Statistics*
- *Patterns and Relationships*
- *Number Sense and Numeration*

Materials: *Newspaper or other sources of weather data, overhead transparency of stem-and-leaf plot*

Cloud cover

NCTM Teaching Standards: *The teacher has a central role in orchestrating the oral and written discourse in ways that contribute to students' understanding of mathematics.*

GETTING STARTED

Pose the following question to students: Why do scientists keep data about the weather by recording such measurements as temperature, precipitation, and barometric pressure?

Accept all answers given by students.

DEVELOPING THE ACTIVITY

Tell students that they will be collecting their own data so that they can summarize their findings and make predictions about the weather in their area. Every day for thirty days, students will collect high and low temperatures.

Class Conversation

- How can we collect weather data? What are sources for weather information? [our own observations and measurements, newspapers, nightly news shows, the Internet, or radio programs]

- How can we make an easy-to-understand record of our data?

- Should everyone collect the same data, or should some groups specialize in different measures?

Encourage suggestions from your students. Students can work in small groups to design a data-recording sheet. You may want to direct all groups to collect high and low daily temperatures. Each group, however, could decide what other data to collect, such as wind direction and speed, times of sunrise and sunset, barometric pressure, and humidity.

Students should record their data for thirty days. You may want to have a regular time during the week to talk about patterns that students see in their data.

After thirty days, have students work in their groups to decide how to display their data. Ask them to share their ideas with the whole class.

Some different ways that students may choose follow:

- A single- or multiple-line graph to show how the selected data change from day to day across the thirty-day period

- A line plot to show the number of days that given conditions were present

Sample Line Plot for Daily High Temperatures

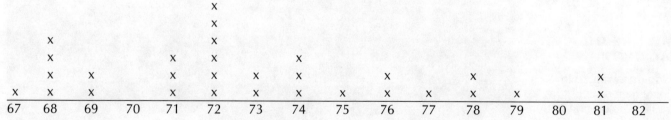

- A stem-and-leaf plot to show frequencies for intervals of temperatures

- A double-bar graph to compare daily high and low temperatures across a given period, such as a week

Class Conversation

- What can you tell me about your graphs?

- If we continue to collect data in the same way, would your graphs look the same next month? Why or why not?
- How will the graphs help us predict weather for the same time next year?
- Why is it important that we try to predict the weather?
- Why would the scientists and engineers at NASA be interested in collecting data about the weather?

Stem-and-Leaf Plots

You may want to use the students' data to introduce or review the use of a stem-and-leaf plot. As you record data in this type of graph, ask students to observe the process and to describe how you have organized the data.

Write the following set of daily high temperatures on an overhead transparency or on chart paper for the students to see. You may want to substitute a student's data set in this graphing activity if one is available.

Note to Primary-Grades Teachers: *Primary-level students can experience the same weather exploration, but perhaps a week, rather than a month, may be more appropriate. You may also want to conduct the activity as a whole-class project.*

Primary-grades students may not have enough place-value experiences necessary to understand the organization of the intervals in the stem-and-leaf plots.

High Temperatures (in degrees Fahrenheit) for 30 Consecutive Days

72 74 72 73 75 74 82 80 82 81

67 64 70 76 73 73 72 76 75 74

80 81 80 82 68 69 76 72 72 77

Stem	Leaves
6	4 7 8 9
7	0 2 2 2 2 2 3 3 3 4 4 4 5 5 6 6 6 7
8	0 0 0 1 1 2 2 2

6 | 4 represents 64.

As you record each leaf value in a row of data, cross off the corresponding data point from the original set of data. For example, in the first row of the stem-and-leaf plot, as you record the 4, cross off the 64 in the given data set. As you record the 7 in the first row, cross off the 67. As you record the 8, cross off the 68. As you record the 9, cross off the 69.

Class Conversation

- What number should I cross off as I record the zero in the second row of our stem-and-leaf plot? As I record each of the five 2s?
- As I record each of the three 3s in the second row, what numbers should I cross off?

Continue with this type of question until students recognize and describe the pattern for the organization.

- In the first row, what does the 6 represent? The 4? The 7? The 8? The 9?
- In the second row, what does the first number, the 7, represent?
- How are the data organized?
- Why do you think that this graph is called a stem-and-leaf plot? What numbers are the stems? What numbers are the leaves?
- How does this graph help us "see" the data better?
- Is this the best way to organize the data? Why or why not?
- What other data could we organize in this way?

Weather Watchers

NCTM Assessment Standards: *As you plan assessment for this activity, consider how the assessment offers opportunities for students to evaluate, reflect on, and improve their own work—to become independent learners. You may want to ask students to—*

- *write a paragraph to summarize their findings from their data and*
- *explain why it is important to keep weather data from year to year.*

CLOSING THE ACTIVITY

Ask students to create their own stem-and-leaf plots using the class data. They can work together in their groups to help one another. Plan time for them to share their data and graphs with their classmates. Ask questions to help them identify patterns and think about predictions for the next week.

FURTHER EXPLORATIONS

- Ask students to keep a journal or record of the predicted weather, using a local newspaper or a news program. Compare that data with data they collect. Are the meteorologists always right? Should we listen to the weather reports before getting ready for school?

- As students become more proficient with the stem-and-leaf plot, you may want to introduce a back-to-back stem-and-leaf plot. For example, students can plot both the high and low temperatures for the thirty days by having a center column for the stems, a left column for the leaves of the low temperatures, and a right column for the leaves of the high temperatures.

- A computer software program can be used to graph the collected data. Ask the class for positive and negative features of using different methods. Was one method easier? Why? Which method was faster? Which type of graph is easier to understand?

- Other important resources are Internet sites related to the study of the weather. See the National Oceanic and Atmospheric Administration (NOAA).

School-Home Connection

Collecting data regularly with their family is a good activity for students to share at home. Remind them to record the source of their data.

APPENDIX A
NASA RESOURCES FOR EDUCATORS

NASA's Central Operation of Resources for Educators (CORE) was established for the national and international distribution of NASA-produced educational materials in audiovisual format. Educators can obtain a catalog and an order form by one of the following methods:

- Write to
  NASA CORE
  Lorain County Joint Vocational School
  15181 Route 58 South
  Oberlin, OH 44074
- Phone (440) 774-1051, ext. 249 or 293
- Fax (440) 774-2144
- E-mail nasaco@leeca.esu.k12.oh.us
- Home Page: http://spacelink.nasa.gov/CORE

Educator Resource Center Network

To make additional information available to the education community, the NASA Education Division has created the NASA Educator Resource Center (ERC) network. ERCs contain a wealth of information for educators: publications, reference books, slide sets, audiocassettes, videotapes, telelecture programs, computer programs, lesson plans, and teacher guides with activities. Educators can preview, copy, or receive NASA materials at these sites. Because each NASA Field Center has its own areas of expertise, no two ERCs are exactly alike. Phone calls are welcome if you are unable to visit the ERC that serves your geographic area. A list of the centers and the regions they serve appears at the right.

Regional Educator Resource Centers (RERCs) offer more educators access to NASA educational materials. NASA has formed partnerships with universities, museums, and other educational institutions to serve as RERCs in many states. A complete list of RERCs is available through CORE or electronically through NASA Spacelink at http://spacelink.nasa.gov.

NASA On-line Resources for Educators provides current educational information and instructional resource materials to teachers, faculty, and students. A wide range of information is available, including science, mathematics, engineering, and technology education lesson plans; historical information related to the aeronautics and space program; current status reports on NASA projects; news releases; information on NASA educational programs; useful software; and graphics files. Educators and students can also use NASA resources as learning tools to explore the Internet, accessing information about educational grants, interacting with other schools that are already online, and participating in online interactive projects by communicating with NASA scientists and engineers and other team members to experience the excitement of real NASA projects.

Access these resources through the NASA Education Home Page: http://www.hq.nasa.gov/education.

Quest is the home of NASA's K–12 Internet Initiative, one of the electronic resources that the agency has developed for the educational community. The project specializes in providing programs, materials, and opportunities for teachers and students to use NASA resources as learning tools to explore the Internet. Through Quest, teachers can access information about educational grants, interact with other schools that are already online, and explore links to other NASA educational resources.

One of Quest's most unusual endeavors is the "Sharing NASA" online interactive project. Students and educators are given the opportunity to communicate with NASA scientists and researchers to experience the excitement of real science in real time. In addition to these programs, the project also houses information

---

AK, AZ, CA, HI, ID, MT, NV, OR, UT, WA, WY
NASA Educator Resource Center
Mail Stop 253-2
NASA Ames Research Center
Moffett Field, CA 94035-1000
Phone: (650) 604-3574

CT, DE, DC, ME, MD, MA, NH, NJ, NY, PA, RI, VT
NASA Educator Resource Laboratory
Mail Code 130.3
NASA Goddard Space Flight Center
Greenbelt, MD 20771-0001
Phone: (301) 286-8570

CO, KS, NE, NM, ND, OK, SD, TX
JSC Educator Resource Center
Space Center Houston
NASA Johnson Space Center
1601 NASA Road One
Houston, TX 77058-3696
Phone: (281) 483-8696

FL, GA, PR, VI
NASA Educator Resource Laboratory
Mail Code ERL
NASA Kennedy Space Center
Kennedy Space Center, FL 32899-0001
Phone: (407) 867-4090

KY, NC, SC, VA, WV
Virginia Air and Space Museum
NASA Educator Resource Center for
  NASA Langley Research Center
600 Settlers Landing Road
Hampton, VA 23669-4033
Phone: (757) 727-0900, ext. 757

IL, IN, MI, MN, OH, WI
NASA Educator Resource Center
Mail Stop 8-1
NASA Lewis Research Center
21000 Brookpark Road
Cleveland, OH 44135-3191
Phone: (216) 433-2017

AL, AR, IA, LA, MO, TN
U.S. Space and Rocket Center
NASA Educator Resource Center for
  NASA Marshall Space Flight Center
P.O. Box 070015
Huntsville, AL 35807-7015
Phone: (205) 544-5812

MS
NASA Educator Resource Center
Building 1200
NASA John C. Stennis Space Center
Stennis Space Center, MS 39529-6000
Phone: (228) 688-3338

NASA Educator Resource Center
JPL Educational Outreach
Mail Stop CS-530
NASA Jet Propulsion Laboratory
4800 Oak Grove Drive
Pasadena, CA 91109-8099
Phone: (818) 354-6916

CA cities near the center
NASA Educator Resource Center for
  NASA Dryden Flight Research Center
45108 North Third Street East
Lancaster, CA 93535
Phone: (805) 948-7347

VA's and MD's Eastern Shores
NASA Educator Resource Lab
Education Complex—Visitor Center
  Building J-1
NASA Wallops Flight Facility
Wallops Island, VA 23337-5099
Phone: (757) 824-2297/2298

about materials that accompany the K–12 Internet Initiative videos. These videos promote the Internet in schools and assist educators in acquiring and integrating the Internet into the classroom (for information about the videotapes, send an e-mail message to video-info@quest.arc.nasa.gov).

Quest can be accessed on the Internet at **http://quest.arc.nasa.gov**. To stay informed about new oppportunities in the Sharing NASA program, send an e-mail message to listmanager@quest.arc.nasa.gov. In the body of the message, write these words: subscribe sharing-nasa. For additional information, send an e-mail message to info@quest.arc.nasa.gov.

NASA Spacelink is one of NASA's electronic resources specifically developed for use by the education community. This comprehensive electronic library contains current and historical information related to NASA's aeronautics and space research. Teachers, faculty, and students will find that Spacelink offers not only information about NASA programs and projects but also teacher guides, pictures, and computer software that can enhance classroom instruction.

Spacelink also provides links to other NASA resources on the Internet. Educators can access materials chosen specifically for their educational value and relevance, including science, mathematics, engineering, and technology education lesson plans; information on NASA educational programs and services; current status reports on agency projects and events; news releases; and television broadcast schedules for NASA Television.

Spacelink can be accessed at the following address: **http://spacelink.nasa.gov**. For additional information, e-mail comments@spacelink.msfc.nasa.gov.

NASA Television (NTV) is the agency's distribution system for live and taped programs. It offers the public a front-row seat for launches and missions, as well as informational and educational programming, historical documentaries, and updates on the latest developments in aeronautics and space science. NTV is transmitted on the GE-2 satellite, transponder 9C at 85 degrees west longitude, vertical polarization, with a frequency of 3880 megahertz and audio of 6.8 megahertz.

Apart from live mission coverage, regular NASA Television programming includes a Video File from noon to 1:00 P.M., a NASA Gallery File from 1:00 to 2:00 P.M., and an Education File from 2:00 to 3:00 P.M. (all times eastern). This sequence is repeated at 3:00 P.M., 6:00 P.M., and 9:00 P.M., Monday through Friday. The NTV Education File features programming for teachers and students on science, mathematics, and technology. NASA Television programming may be videotaped for later use.

For more information on NASA Television, contact NASA Headquarters, Code P-2, NASA TV, Washington, DC 20546-0001; phone: (202) 358-3572; NTV Home Page: **http://www.hq.nasa.gov/ntv.html**.

The brochure *How to Access NASA's Education Materials and Services,* EP-1996-11-345-HQ, serves as a guide to accessing a variety of NASA materials and services for educators. Copies are available through the ERC network or electronically through NASA Spacelink. NASA Spacelink can be accessed at the following address: **http://spacelink.nasa.gov**.

APPENDIX B
CHARTING THE PLANETS

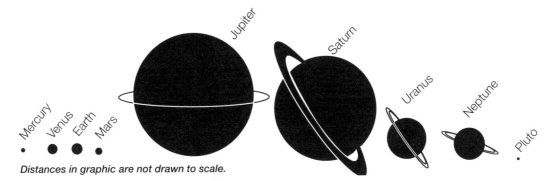

Distances in graphic are not drawn to scale.

Categories	Mercury	Venus	Earth	Mars	Jupiter	Saturn	Uranus	Neptune	Pluto
1. Mean Distance from Sun (millions of kilometers)	57.9	108.2	149.6	227.9	778.3	1 427	2 871	4 497	5 913
2. Period of Revolution	87.9 days	224.7 days	365.3 days	687.0 days	11.86 years	29.46 years	84.0 years	165.0 years	248.0 years
3. Equatorial Diameter (kilometers)	4 880	12 100	12 756	6 786.8	143 200	120 000	51 800	49 528	~2 330
4. Atmosphere (main components)	Essentially none	Carbon dioxide	Nitrogen Oxygen	Carbon dioxide	Hydrogen Helium	Hydrogen Helium	Helium Hydrogen Methane	Hydrogen Helium Methane	Methane + ?
5. Moons	0	0	1	2	16	18+ (?)	15	8	1
6. Rings	0	0	0	0	1	1000 (?)	11	4	0
7. Inclination of Orbit to Ecliptic	7°	3.4°	0°	1.85°	1.3°	2.5°	0.8°	1.8°	17.1°
8. Eccentricity of Orbit	0.206	0.007	0.017	0.093	0.048	0.056	0.046	0.009	0.248
9. Rotation Period	58.9 days	243 days retrograde	23 hours 56 min	24 hours 37 min	9 hours 55 min	10 hours 40 min	17 hours 12 min retrograde	16 hours 7 min	6 days 9 hours 18 min retrograde
10. Inclination of Axis*	0.0°	177.2°	23°27'	25°12'	3°5'	26°4'	97°55'	28°48'	120°

* Inclinations greater than 90° imply retrograde rotation.

BIBLIOGRAPHY

Professional References

National Council of Teachers of Mathematics (NCTM). *Assessment Standards for School Mathematics*. Reston, Va.: NCTM, 1995.

———. *Curriculum and Evaluation Standards for School Mathematics*. Reston, Va.: NCTM, 1989.

———. *Professional Standards for Teaching Mathematics*. Reston, Va.: NCTM, 1991.

Student References

Ardley, Neil. *The Science Book of Air*. San Diego, Calif.: Harcourt Brace Jovanovich, 1991.

Armbruster, Ann, and Elizabeth A. Taylor. *Astronaut Training*. New York: Watts, 1990.

Beasant, Pam. *1000 Facts about Space*. New York: Scholastic, 1992.

Bellville, Cheryl Walsh. *The Airplane Book*. Minneapolis: Carolrhoda Books, 1991.

Berliner, Don. *Before the Wright Brothers*. Minneapolis: Lerner Publications, 1990.

———. *Distance Flights*. Minneapolis: Lerner Publications, 1990.

Blackburn, Ken, and Jeff Lammers. *The World Record Paper Airplane Book*. New York: Workman, 1994.

Booth, Nicholas. *The Encyclopedia of Space*. New York: Mallard Press, 1990.

Boyne, Walter, Jr. *The Smithsonian Book of Flight for Young People*. New York: Atheneum, 1988.

Briggs, Carole S. *Research Balloons: Exploring Hidden Worlds*. Minneapolis: Lerner Publications, 1988.

———. *Women in Aviation*. Minneapolis: Lerner Publications, 1990.

Brigman, Chris, et al. *How to Fly the Space Shuttle*. Sante Fe, N.M.: J. Muir Publications, 1992.

Burch, J. *Astronauts*. Ada, Okla.: Garrett, 1992.

Burns, K., and W. Miles. *Black Stars in Orbit: NASA's African-American Astronauts*. San Diego, Calif.: Harcourt Brace, 1995.

Churchill, E. R., and J. Michaels. *Fabulous Paper Airplanes*. New York: Sterling, 1992.

———. *Fantastic Paper Flying Machines*. New York: Sterling, 1994.

Cole, Joanna. *Magic Schoolbus Lost in the Solar System*. New York: Scholastic, 1992.

Couper, Heather, and Nigel Henbest. *How the Universe Works*. Pleasantville, N.Y.: Reader's Digest Association, 1994.

Darling, David J. *Up, Up, and Away*. New York: Dillon, 1991.

Dixon, M., and S. Wheele. *Flight*. New York: Bookwright Press, 1991.

Doherty, Paul, Don Rathjen, and Exploratorium Teacher Institute Staff. *The Spinning Blackboard and Other Dynamic Experiments on Force and Motion*. New York: John Wiley & Sons, 1996.

Hardesty, Von, and Dominick Pisano. *Black Wings: The American Black in Aviation*. Washington, D.C.: Smithsonian Institution Press, 1984.

Joels, Kerry M., et al. *The Space Shuttle Operator's Manual*. New York: Ballantine Books, 1988.

Johnstone, Michael. *Airplanes*. New York: Dorling Kindersley, 1994.

Kerrod, Robin. *The Children's Space Atlas: A Voyage of Discovery for Young Astronauts*. Brookfield, Conn.: Millbrook Press, 1992.

Lopez, Donald. *Flight*. Alexandria, Va.: Time-Life Books, 1995.

Maurer, Richard. *Rocket! How a Toy Launched the Space Age*. New York: Crown, 1995.

McKeeve, Michael, and Georganne Irvine. *A Day in the Life of a Test Pilot*. Mahwah, N.J.: Troll Communications, 1991.

Michael, David, and David Jefferies. *Making Kites*. New York: Larousse Kingfisher Chambers, 1993.

Moeschl, Richard. *Exploring the Sky: 100 Projects for Beginning Astronomers*. Chicago: Chicago Review, 1993.

Moser, Barry. *Fly! A Brief History of Flight Illustrated*. New York: Harper-Collins, 1993.

Murray, P. *The Space Shuttle*. Mankato, Minn.: Child's World, 1994.

Nahum, Andrew. *Flying Machine*. New York: Knopf, 1990.

Nicolson, I., and S. Quigley. *Explore the World of Space and the Universe*. Racine, Wisc.: Western Publishing Co., 1992.

Parker, Steve. *Airplanes*. Brookfield, Conn.: Millbrook Press, 1995.

———. *Flight and Flying Machines*. New York: Dorling Kindersley, 1990.

———. *What's Inside Airplanes?* New York: Peter Bedrick Books, 1993.

Rupp, Rebecca. *Everything You Never Learned about Birds*. Pownal, Vt.: Storey Communications, 1995.

Schmidt, N. *Super Paper Airplanes: Biplanes to Space Planes*. New York: Sterling, 1995.

Scholastic. *The Story of Flight*. New York: Scholastic, 1995.

Schwartz, David M. *How Much Is a Million?* New York: Scholastic, 1986.

Scott, E., and M. Miller. *Adventure in Space: The Flight to Fix the Hubble*. New York: Hyperion Books for Children, 1990.

Skurzynski, Gloria. *Almost the Real Thing: Simulation in Your High-Tech World*. New York: Macmillan Child Group, 1991.

Solomon, Maury. *An Album of Voyager*. New York: Franklin Watts, 1990.

Stott, Carole. *Space*. New York: Watts Books, 1995.

Sullivan, George. *The Day We Walked on the Moon: A Photo History of Space Exploration*. New York: Scholastic, 1992.

Taylor, Beverley, James Poth, and Dwight Portman. *Teaching Physics with Toys: Activities for Grades K–9*. New York: McGraw-Hill, 1996.

Vancleave, Janice P. *Janice Vancleave's Astronomy for Every Kid: 101 Easy Experiments That Really Work*. New York: John Wiley & Sons, 1991.

Verdet, J. *Earth, Sky, and Beyond: A Journey through Space*. New York: Lodestar Books, 1995.

Vogt, Gregory. *The Hubble Space Telescope*. Brookfield, Conn.: Millbrook Press, 1992.

Ward, A. *Sky and Weather*. New York: Franklin Watts, 1992.

Wellington, J., and F. Lloyd. *The Super Science Book of Space*. New York: Thomson Learning, 1993.

Wells, Robert E. *Is a Blue Whale the Biggest Thing There Is?* Morton Grove, Ill.: A. Whitman, 1993.

Williams, J., and M. S. Walker. *Projects with Flight*. Milwaukee, Wisc.: Gareth Stevens Children's Books, 1992.

Yenne, Bill. *The Astronauts: The First 25 Years of Manned Space Flight*. New York: Hamlyn/Bison, 1986.